T0199344

Rocks and Minerals in Thin Section

Second Edition

W.S. MacKenzie
Emeritus Professor of Petrology,
University of Manchester, England

A.E. Adams
Senior Lecturer in Geology (retired),
University of Manchester, England

K.H. Brodie
Senior Lecturer in Geology,
University of Manchester, England

CRC Press
Taylor & Francis Group
Boca Raton London New York Leiden

CRC Press is an imprint of the
Taylor & Francis Group, an **informa** business

A BALKEMA BOOK

Revised 2nd edition

CRC Press/Balkema is an imprint of the Taylor & Francis Group, an informa business

© 2017 Taylor & Francis Group, London, UK

Typeset by MPS Limited, Chennai, India

Printed and bound in India by Replika Press Private Limited

Published by: CRC Press/Balkema
                Schipholweg 107C, 2316 XC Leiden, The Netherlands
                e-mail: Pub.NL@taylorandfrancis.com
                www.crcpress.com – www.taylorandfrancis.com

ISBN: 978-1-138-02806-7 (Pbk)
ISBN: 978-1-138-09184-9 (Hbk)
ISBN: 978-1-315-11636-5 (eBook)

Contents

Preface

This atlas has been prepared for students of earth science, geology, mineralogy and physical geography who require a text for practical classes on rocks and minerals under the microscope. While the book's prime purpose is as an introduction to the subject for college and university students as an essential part of their course, we hope that amateur geologists and mineralogists will also find it useful and attractive.

We have tried to make the text and pictures self-contained such that an individual who has access to a polarizing microscope and a collection of thin sections of rocks can begin recognizing minerals and naming rocks without supervision. Our aim has been to provide a manual for use in practical classes by showing illustrations of some of the diagnostic properties of minerals and introducing the most common rock-forming minerals. We then illustrate a representative selection of igneous, sedimentary and metamorphic rocks.

We have deliberately limited the scope of the introduction to optical mineralogy and have assumed little knowledge of crystallography or physical optics. We would hope, however, that the coverage will encourage students to study the elements of crystal symmetry and thus be in a better position to understand crystal optics. This would assist the student to progress to the use of optical techniques not covered here, such as the use of convergent light.

Most of the photographs of rocks in this book have been taken at low magnification to illustrate representative views of the constituent minerals and their interrelationships. The photographs were taken either in plane-polarized light or under crossed polars: in many cases the same field of view is shown under both conditions. Some of the photographs reproduced here are from thin sections which have been used previously for other publications. However, all the photographs here were made especially for this publication as 6×9 cm transparencies. We have had the advantage of having access to a Zeiss Ultraphot microscope for this purpose.

Acknowledgements

Most of the thin sections illustrated are from the teaching collections of Manchester University Geology Department, and we are indebted to our colleagues who have collected these specimens over many years. We are especially grateful to those who have supplied us with additional material, particularly Giles Droop, Alistair Gray and John Wadsworth.

Colin Donaldson kindly agreed to read the first two sections and made useful comments on the text. The authors alone, however, are responsible for the choice of rock types and for their descriptions.

We also record our grateful thanks to Carolyn Holloway for her typing of the text and for her patience during all our changes of mind.

About the authors

Professor William MacKenzie's main research area was the application of high pressure-temperature experiments to petrology, with a particular interest in properties of high temperature alkali feldspars. He established a world-renowned experimental petrology laboratory at Manchester in 1956. His interest and expertise in microscopy and photography combined with his enthusiasm for teaching, led to the publication of a series of Atlases of rock forming minerals. Many of his original plates have been retained in this new edition.

Dr Anthony Adams was a popular lecturer in sedimentology at the University of Manchester for much of his career until his retirement. His research was focused on carbonate sedimentology, with a special interest in carbonate rocks of south Wales and the western Mediterranean.

Dr Katharine Brodie has taught petrology for the last 40 years, first at Imperial College, London and subsequently at the University of Manchester. Her main research interest has been the interactions between metamorphism and deformation, linking laboratory experimental work with natural examples, on scales ranging from microscopic to larger scale tectonic processes.

Second edition

In producing a second edition, many of the original micrographs have been retained but the classification of igneous and of metamorphic rocks has been updated to reflect current usage. Mudstones and porosity has been added to the sedimentary section and the metamorphic section has been expanded. The new photographs have been taken using a Nikon XX microscope.

Giles Droop, Alison Pawley and Brian O'Driscoll are thanked for providing additional samples and comments on the revised version.

Introduction

To gain an introduction to the identification of minerals and rocks under the polarizing microscope, the student has first to acquire some knowledge of the compound microscope in its simplest form. This may sound like a contradiction in terms but a magnifying lens is correctly described as a simple microscope, whereas a compound microscope has at least two lenses, one producing a real image of the object (the objective lens) and the other magnifying this image (the eyepiece). Magnifications greater than 20 times are normally obtained using a compound microscope. It is assumed that the operations of focusing the microscope, adjusting the illumination and ensuring that the centring of the stage with respect to the optic axis of the microscope can all be accomplished. It is also assumed that the student has access to a collection of thin sections of rocks ground to the standard thickness of one thousandth of an inch or 0.03 mm.

First we hope to help the student to describe *minerals*. After only a few hours study the beginner will guess the identity of some minerals as he or she becomes familiar with the appearance of the commonest minerals under the microscope by observation of their properties. This can work satisfactorily as long as the student can describe the optical properties correctly–if one or more of the properties do not correspond to the mineral suggested then the identification is incorrect and he or she must think again.

Rocks are composed of aggregates of minerals. After determining the minerals, the identification of a rock depends on the relative abundance of the minerals and on the textural relationships between them. No attempt has been made to introduce the student to petrogenesis, i.e., the study of the origins of rocks. Our aim is to introduce the subject of *petrography* or the description of rocks, since it is extremely important to distinguish observations from hypotheses, and the observations must come first. However, some simple assumptions about the origins of rocks must be made before they can be classified, but these are unlikely to be controversial. A short account of the nomenclature of the rocks is given at the beginning of each section. A brief introduction to textural observations is provided as more detailed observations and interpretations are beyond the scope of this book.

We have not given a complete petrographic description of any rock because this can only be written after examination of an actual thin section of rock viewed at different magnifications and covering an area representative of the whole rock.

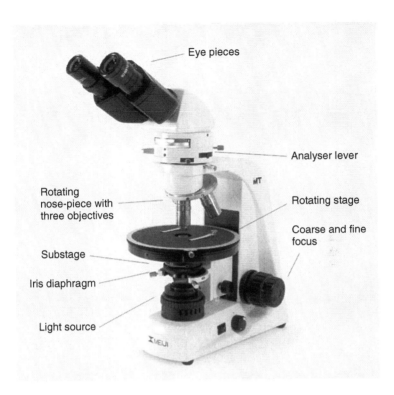

Eye pieces

Analyser lever

Rotating nose-piece with three objectives

Rotating stage

Coarse and fine focus

Substage

Iris diaphragm

Light source

A student-model petrographic microscope.

PART 1

Optical mineralogy

The polarizing microscope

The polarizing or petrographic microscope is distinguished from the more usual biological microscope in that it is equipped with a rotating stage and two polarizing filters, one below the stage and the other above it. Ordinary light may be considered to consist of waves vibrating in all directions whereas polarized light consists of vibrations in one plane only—the plane of polarization. The polarization filters are made from material known as polaroid. Polaroid is used in some makes of sun glasses and photographic filters to cut out glare from reflecting surfaces. The polarizing filters in the microscope are normally set so that their polarization directions are at right angles to one another and parallel to the cross-wires in the eyepiece of the microscope. The polarizing filter below the stage is known as the polarizer, that above the stage is the analyser. The analyser is mounted in such a way that it can be removed from the light path so that the rock section can be studied in plane-polarized light. When the analyser is inserted the sample is said to be observed with crossed polars. When there is no thin section on the microscope stage no light can be seen on looking down the microscope when the polars are crossed because the polarized light emerging from the polarizer is blocked by the analyser.

A polarizing microscope is illustrated opposite. This model, produced by Meiji, mainly for student use, has all the facilities required for petrographic study of rock sections in transmitted light. The parts of the microscope with which the beginner must become familiar are marked on the photograph. This version has one eye pieces (monocular) but binocular versions are commonly used.

This instrument has a nosepiece carrying three objectives each having a different magnification: rotation of the nosepiece permits a change in magnification by bringing one objective into a vertical position directly above the thin section. The objectives are designed to be parfocal: thus when an objective is changed only a small adjustment in focus is necessary.

Focusing a microscope involves adjusting the distance between the objective and the object being examined. In this instrument focusing is achieved by altering the height of the stage and the focusing controls can be seen at the lower end at the back of the microscope.

The substage assembly carries, in addition to the lower polarizer, a condensing lens and an iris diaphragm. These facilities permit observation of minerals in a strongly converging beam of polarized light, as well as in a non-converging (i.e. parallel) beam. Examination of minerals in convergent light is beyond the scope of this text. The iris diaphragm is also used in restricting the aperture:

- To obtain improved contrast between minerals of slightly different refractive indices.
- For observing the Becke Line (see page 12) to determine the relative refractive indices of adjacent minerals or a mineral and the mounting material.

An inexpensive biological microscope can be obtained much more readily than a polarizing microscope and by incorporating two pieces of polaroid in the light path such a microscope may be used for the study of thin sections of rocks provided that the polaroid above the thin section can be easily removed and re-inserted. The facility of rotating the microscope stage will not normally be available in a biological microscope and in such a case it would be necessary to be able to rotate the lower polarizer. There are two reasons why it is desirable to be able to rotate the stage or the lower polarizer. These are:

- To observe pleochroism (i.e. change in colour of a mineral as seen in plane-polarized light, when the mineral is rotated with respect to the plane of polarization of light.)
- To measure extinction angles (see page 18).

Description of minerals

To describe a mineral and so identify it correctly a student must be able to:

- Describe the shape of the crystals.
- Note their colour and any change in colour on rotation of the stage in plane-polarized light.
- Note the presence of one or more cleavages.
- Recognize differences in refractive index of transparent minerals and determine which has the higher refractive index of two adjacent minerals.
- Observe the interference colour with crossed polars and identify the maximum interference colour.
- Note the relationship between the extinction position and any cleavages or traces of crystal faces.
- Observe any twinning or zoning of the crystals.

These properties are treated in some detail below and are illustrated where possible.

Shape and habit of crystals

In a completely crystalline rock it is unlikely that the faces of all the crystals will be well-developed because they interfere with one another during growth. In an igneous rock the first crystals to grow are likely to have well formed crystal faces because they have probably grown freely in a liquid. In some metamorphic and sedimentary rocks, crystals with well-developed crystal faces are presumed to have grown in an environment consisting mainly of solids but with the possibility of fluid in the interstices.

Crystals whose outlines in thin section show well defined straight edges, which are slices through the faces of the crystal, are described as *euhedral* crystals (1, 2);

1   Euhedral crystals (porphyroblasts) of garnet in a metamorphic rock.

2   Euhedral crystals of nepheline in an igneous rock.

crystals which have no recognizable straight edges are *anhedral* and crystals with some straight edges and others curved are *subhedral*.

In an igneous rock, large crystals in a matrix or groundmass of much smaller crystals are described as *phenocrysts* (**3**). In a metamorphic rock similar large crystals which have grown in a mass of smaller crystals are termed *porphyroblasts* (**1, 4**). If the porphyroblast contains abundant small inclusions, it is termed a *poikiloblast*. In some rocks it is not certain whether the large crystals grew from an igneous magma or in a later stage metamorphic event. In these cases it is perhaps better to describe the crystals as *megacrysts*. In some deformed metamorphic rocks, large relict grains are preserved in a finer grained matrix and these are termed *porphyroclasts*.

To describe the outlines of crystals as seen in thin section, such words as rectangular, square, hexagonal, diamond-shaped, or rounded are self-explanatory.

The term *habit* is used to indicate the shape of crystals as seen in hand specimen or deduced from several cross-sections in a thin slice. The following terms are used: *needle-shaped*, (or *acicular*), *prismatic* and *tabular*. The first of these terms is self-explanatory (**5**). Prismatic is the term used to describe crystals which have similar dimensions in two directions and are elongated in the third dimension (**6**). Tabular habit is used to describe crystals which are flat in one plane.

A mineral may be characterized by a particular habit but in some rocks one mineral may display two different habits.

3

1 mm

3 Phenocrysts of olivine in an igneous rock.

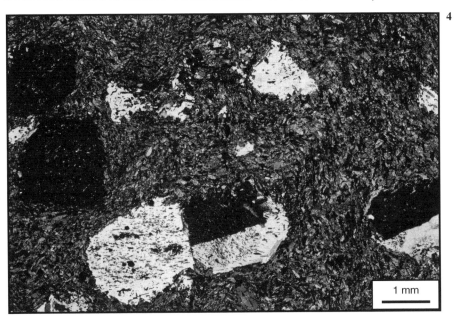

4 Porphyroblasts of albite in a metamorphic rock.

5 Needle-shaped or acicular crystals of tourmaline.

Colour and pleochroism

Many minerals, although coloured in hand specimen, may be nearly colourless in thin section. A few common minerals are easily recognized by their colour in thin section, e.g., the mineral biotite is usually brown (**8**). Some minerals are opaque in thin section and their properties can only be studied with a reflected light microscope. A mineral which is coloured in thin section may show a different colour or shades of one colour as the microscope stage is rotated. Because crystals in a rock are usually randomly arranged and hence cut in different directions in a thin section, they are likely to show different colours or shades of one colour in a section. The colour of a mineral when observed in plane-polarized light is termed its *absorption colour* and the phenomenon of variation in colour depending on the orientation of a crystal with respect to the plane of polarization of the light is known as *pleochroism* (**7, 8**). This is a very useful diagnostic property for some minerals.

6

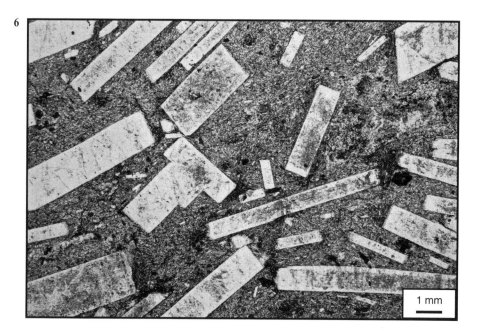

1 mm

6 Prismatic crystals of sanidine.

7 & 8 In **7**, olive-green crystals of tourmaline, a complex boron aluminium silicate, are intergrown with pale yellow biotite. In **8**, taken after rotating the polarizer through 90°, many of the tourmaline crystals have changed and are colourless and much of the biotite is brown. The orientation of the polarizer is shown by the double headed arrow beside each figure. The extent to which the crystals change colour depends on their orientation.

7

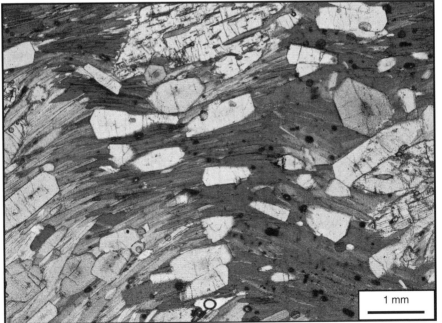

8

Cleavage

Many minerals break or cleave along certain planes, the positions of which are controlled by the atomic structure of the minerals. Between cleavage planes the atomic bonding is weak compared to that within the planes. The presence or absence of cleavage and the angles between cleavages if more than one cleavage is present, may be of diagnostic value.

Crystals of mica can be easily separated into thin sheets because micas have a perfect cleavage in one plane. In crystals cut at right angles to the cleavage plane, the cleavage is visible in thin section as a set of thin straight, parallel, dark lines, (**9**) whereas if the crystal is cut nearly parallel to the cleavage it is not visible. Some minerals cleave parallel to more than one plane and the angle between two cleavages can be diagnostic of certain minerals; thus in the pyroxene group of minerals two cleavages are at 90° (**10**), whereas in the amphiboles the cleavages intersect at an angle of 120° (**11**). In a thin section the angle between two cleavages can only be measured with accuracy when the thin section is cut at, or nearly at, right angles to both cleavages. Cleavages tend to be parallel to crystal faces although this is not always the case. In **10** and **11** crystal boundaries are parallel to both cleavages in the pyroxene and the amphibole.

9

9 Brown biotite mica with one cleavage and colourless, high relief kyanite with two cleavages at ~80°.

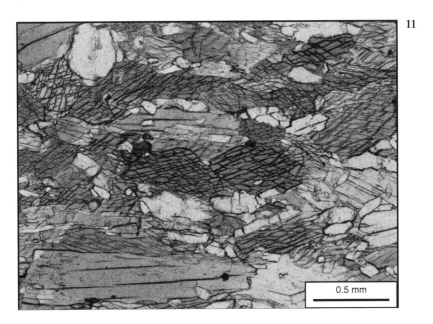

10 Clinopyroxene crystals showing two cleavages at approximately 90°. There are crystal faces parallel to both cleavages.

11 Amphibole crystals with light brown coloured (pleochroism) grains showing two cleavages at approximately 120°. The blue/green crystals are also amphibole in a different orientation showing a single cleavage.

Relief

Colourless minerals of similar refractive index and having refractive indices close to that of the mounting medium do not show distinct boundaries when seen under the microscope. The greater the difference between the refractive index of a mineral and its surrounding material the greater its *relief* (**12, 13**). When differences in the refractive index are small it is necessary to partially close the sub-stage diaphragm to detect differences in relief and if the microscope is not equipped with a substage diaphragm it may be difficult or impossible to detect differences in refractive indices or relief (see discussion of Becke line below).

Minerals have one, two or three refractive indices, depending on their symmetry. On viewing a mineral in thin section in polarized light, its relief may change when rotating the microscope stage since the refractive index of the mineral which is being compared with the mounting medium may change. A few minerals have very large differences between their maximum and minimum refractive index and in such cases the change in relief may be considerable; this is known as *twinkling* and is characteristic of the carbonate minerals (**14, 15**).

12
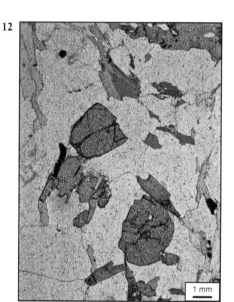

13

12 Crystals having higher refractive indices than others stand out in relief against the background which is mainly quartz. The two minerals showing very high relief in this figure are kyanite and garnet; the brown mineral is biotite and shows moderate relief against the colourless quartz.

13 The elongated crystals shown here are of corundum (Al_2O_3). They have much higher refractive indices than the feldspar in which they are embedded and hence stand out in relief.

0.5 mm

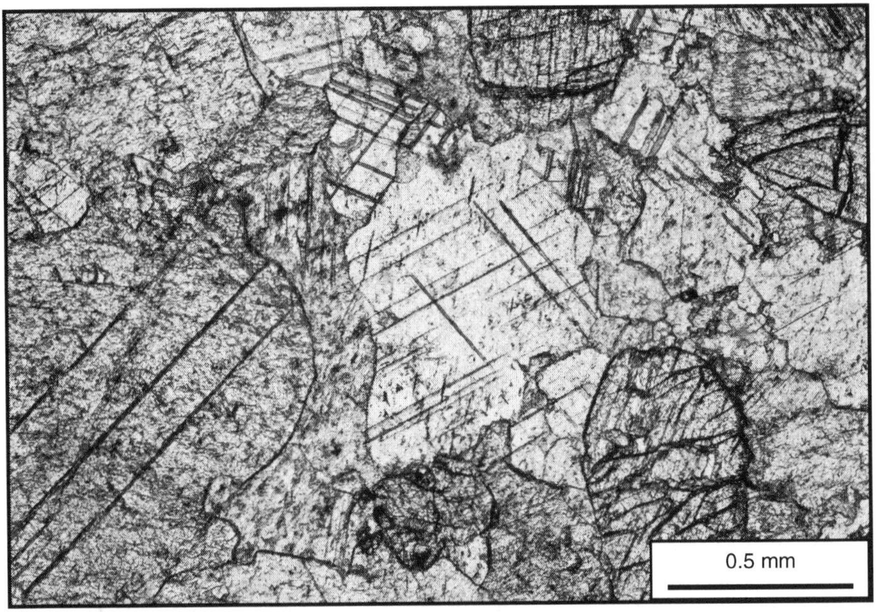

0.5 mm

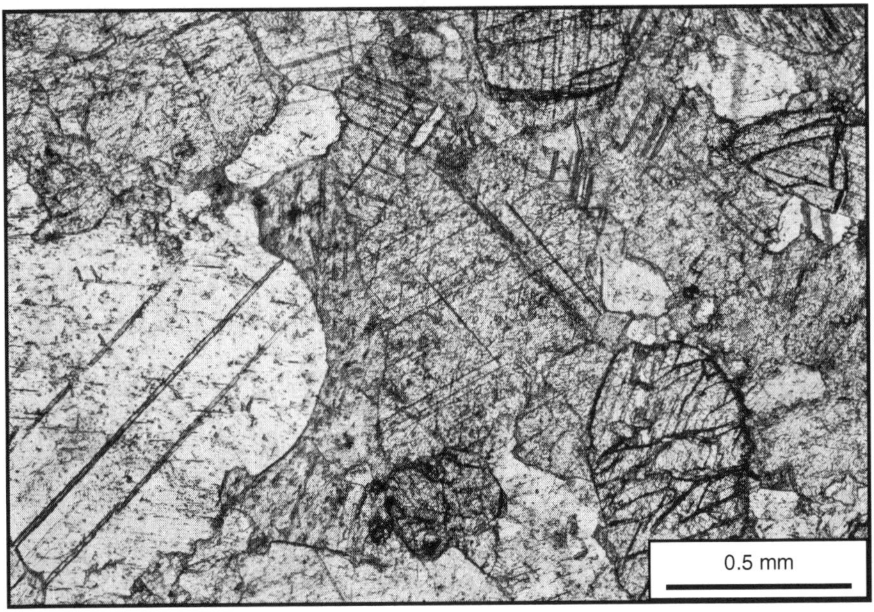

14 & 15 These two figures show calcite crystals in a marble. The orientation of the polarizer is shown by the double headed arrows adjacent to the figures and we can see that the relief of each of the calcite crystals relative to its neighbouring crystals changes with rotation of the polarizer.

In attempting to identify minerals, it is often desirable to know which of two adjacent materials has the higher refractive index. The boundary between materials of differing refractive index is characterized by a bright line which can be enhanced by partially closing the sub-stage diaphragm and defocusing the image slightly; this bright line is known as the *Becke line*. If the tube of the microscope is raised or the stage lowered (depending on the method of focusing), it is observed that the Becke line moves into the material which has the higher refractive index and on lowering the tube, or raising the stage, the bright line moves into the lower refractive index material (**16–18**).

If instead of a bright line the boundary between two minerals is marked by a faint blue and yellow fringe this is an indication that the two minerals have very similar refractive indices and only in the light of a given colour or wavelength could an observer specify which mineral has the higher index of refraction.

16

1 mm

16–18   *The Becke line*: In **16** the right hand side of the field of view is occupied by a few crystals of muscovite whereas the left side is the mounting medium. This figure was taken with the analyser inserted: the mounting medium appears black since it is isotropic (see page 24) whereas the muscovite shows bright interference colours. In **17** and **18** the analyser has been removed and the same field of view can be seen in plane-polarized light. To compare the refractive index of muscovite with that of the mounting medium it is necessary to defocus the microscope—in **17** the microscope tube has been lowered below the position of sharp focus and in **18** the tube has been raised above the position of sharp focus. The bright line, which marks the boundary between the muscovite and the mounting medium, can be seen to have transferred from within the mounting medium in **16** into the muscovite in **17**—the rule is: on raising the microscope tube the Becke line moves into the material of higher refractive index. Thus it can be seen that muscovite has a higher refractive index than the mounting medium.

17

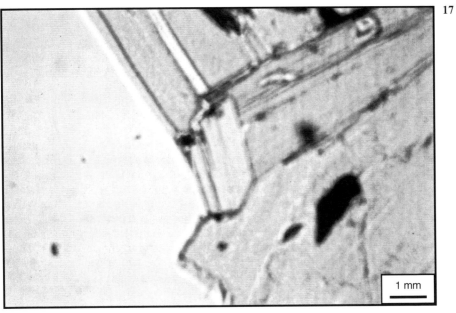

1 mm

18

1 mm

Birefringence

Although values of refractive indices of minerals are of great diagnostic value, it is very difficult to measure them accurately, especially in the case of minerals which have three refractive indices and when the indices are greater than $1\cdot70$. Most mineralogists know how to measure a refractive index using liquids of known refractive index, but very rarely do so except in the case of a new mineral where it is necessary to report its physical constants. Minerals which have more than one refractive index have a property which is known as *double refraction*. A quantitative measure of double refraction is *birefringence,* defined as the difference between the maximum and minimum refractive indices of a mineral. Birefringence can be measured fairly readily and with considerable accuracy.

When polarized light enters most crystals, it is split into two components each having a different velocity; the two light waves become out of phase as they travel through the crystal because of their differing velocities. On emerging from the mineral the two rays interfere with one another and, when observed with the analyser inserted in the light path, show what are known as *interference colours*. These colours are similar to those seen when a thin film of oil is observed on a wet street.

The interference colours shown by a mineral in thin section chiefly depend on three factors:

- the birefringence of the mineral,
- the thickness of the section,
- the orientation in which the mineral is cut.

The second variable is eliminated by cutting all rock sections to a standard thickness of 0.03 mm. To allow for differences in orientation and so eliminate the third variable only the *maximum* value of the interference colour is considered and the value of the birefringence is obtained from the accompanying chart (**19**). This *birefringence chart* shows the interference colours in a section of standard thickness of a colourless mineral corresponding with the value of its birefringence. The common minerals illustrated in this book are indicated at the appropriate birefringence value.

The low interference colours are grey and white and these are at the top of the chart. The chart is divided into *orders*; the first three orders are shown. Most common minerals are covered by the range of birefringence shown, except for the carbonates in which the birefringence is nearly 0.18. The high-order colours shown by the carbonates are illustrated in **71.**

19 *Birefringence Chart.* Birefringence of Michel-Levy interference colours (0.03mm thin section) for some common rock-forming minerals. The chart was kindly provided by Prof. Takenori Kato, Nagoya University, Japan, produced by a programme to synthesize interference colour (Kato 2001), http://www.nendai.nagoya-u.ac.jp.

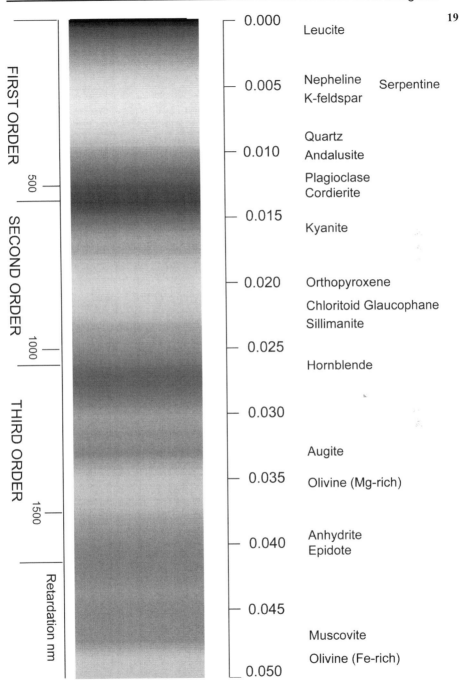

Birefringence

19

FIRST ORDER

500

SECOND ORDER

1000

THIRD ORDER

1500

Retardation nm

0.000 — Leucite

0.005 — Nepheline   Serpentine
K-feldspar

0.010 — Quartz
Andalusite

Plagioclase
Cordierite

0.015 — Kyanite

0.020 — Orthopyroxene

Chloritoid Glaucophane
Sillimanite

0.025 —

Hornblende

0.030 —

Augite

0.035 — Olivine (Mg-rich)

0.040 — Anhydrite
Epidote

0.045 —

Muscovite

Olivine (Fe-rich)

0.050 —

15

A single crystal of a mineral may show any colour between that corresponding to its maximum birefringence colour and black corresponding to zero birefringence, depending on the orientation of the crystal. For a given mineral in a thin section of standard thickness only the maximum colour is of diagnostic value and defines the birefringence (20).

Some minerals show interference colours which are not represented on the birefringence chart. These colours are shades of blue, yellow or brown and are known as *anomalous* colours. If the birefringence of a mineral varies appreciably with the wavelength of light, some colours may be reduced in intensity and so the resultant interference colours are anomalous. If the absorption colour of a mineral is strong, it may affect the interference colour and thus also produce an anomalous colour. A few common minerals are characterized by anomalous interference colours and this may help in identification—e.g. chlorite (52).

It was noted above that minerals may have one, two or three refractive indices. Those which have only one refractive index have structures made up of very regular arrangements of atoms so that light passes through a crystal with the same velocity irrespective of the direction in which it travels. Such minerals show no double refraction and appear black when viewed with crossed polars: these minerals are said to be *isotropic*.

Materials such as glass and liquids are also isotropic but for a very different reason: they are isotropic because they usually have a very disordered arrangement of atoms and in consequence light passes through such materials with the same velocity irrespective of its direction. The mounting materials used for making thin sections are isotropic (and this can be useful to check that the polars on the microscope are correctly aligned as the field of view should be black with the polars crossed).

Minerals which have two refractive indices possess one unique direction in which they show no double refraction and minerals which have three refractive indices have two directions in which they show no double refraction and so appear black when observed between crossed polars. In a thin section the proportion of crystals which have been cut exactly at right angles to one of these directions is small but for more advanced optical techniques it may be desirable to look for such sections.

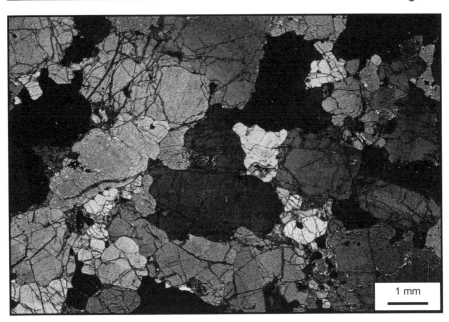

20   This view is of a rock consisting of a number of crystals of the same mineral showing a range of interference colours when viewed between crossed polars. A few crystals show grey or white first-order interference colours, one large crystal to the left of centre of the field of view shows a first-order red colour. The crystal just below the centre of the field of view shows a blue colour and below that the green colour could be a third-order colour. Thus the birefringence of this mineral on the basis of the highest order colour seen in this view is about 0.040, provided that the section is of the correct thickness. The rock is a dunite which is a monomineralic rock consisting almost entirely of olivine.

Extinction angles

The interference colour of each mineral grain in thin section, observed with crossed polars, changes in intensity as the stage is rotated and the intensity falls to zero at every 90° of rotation (i.e. no light is seen by the observer from this crystal). The positions in which a particular grain is black are known as the *extinction positions* for that crystal. The angle between an extinction position and some well defined direction in a crystal is known as the *extinction angle* for that crystal: it is usually quoted as less than 45°, although sometimes the complementary angle is given. Since an extinction angle for a given orientation of a crystal or a maximum extinction angle, obtained by measurements from a number of crystals of the same mineral, may be of diagnostic value, the method of measuring an extinction angle is described briefly below and illustrated in **21–23**.

The thin section should be held in place by one of the spring clips on the microscope stage. Either a straight edge, representing a crystal face, or a cleavage direction of the crystal being studied is set parallel to one of the cross-hairs in the eyepiece and the angular position of the stage read from the fixed vernier scale. This should be done with the analyser removed from the light path. The analyser is then inserted and the stage rotated slowly to one of the extinction positions and the angular position read from the vernier scale. The difference in the two readings is the extinction angle for this particular crystal. If the angle is zero the crystal has *straight extinction*—non zero values are described as *oblique extinction*. An extinction position which bisects the angle between two cleavages is known as *symmetrical extinction*.

21–23   The main part of the field of view is occupied by a crystal of kyanite; a cleavage has been set parallel to the long edge of **21**. The interference colour shown by the kyanite is a first-order pale yellow. In **22** the microscope stage has been rotated through 15° and the brightness of the interference colour has become less intense. In **23** the microscope stage has been rotated through 30° and here the mineral is completely black—it is in the extinction position and only the inclusions of other minerals show interference colours. In this orientation the extinction angle of kyanite is 30°—a value which is characteristic of this mineral when measured from the cleavage shown here.

21

0.5 mm

22

0.5 mm

23

0.5 mm

Twinning and zoning

Many minerals occur in what are known as *twins*. Twins are crystals of the same mineral in which the orientations of the two or more parts have a simple relationship to each other, e.g. a rotation through 180° around one of the crystallographic axes, or reflection across a plane in the crystal (**24**). When this twin operation is repeated a number of times the crystals are described as *polysynthetically twinned* or as showing *multiple twinning*: in this case alternate lamellae show the same orientation.

The commonest rock-forming minerals in the earth's crust are the feldspars and certain types of twinning are characteristic of the different feldspars. The sodium-calcium or plagioclase feldspars invariably show polysynthetic twinning and an estimate of the sodium to calcium ratio may sometimes be obtained from a measurement of the extinction angle or of the maximum extinction angle depending on the orientation of the crystals. In Part 2 we describe a method of determining the sodium to calcium ratio of the plagioclase feldspars from extinction angle measurements of twinned crystals.

24

1 mm

24 *Twinning*. This figure shows a few crystals of pyroxene taken with polars crossed. Some of the crystals have a dividing line and a change of interference colour across this line: this is due to twinning. If the crystal consists of only two parts it is simply twinned. Very often two different orientations are intergrown so that alternate lamellae have different orientations and so different interference colours.

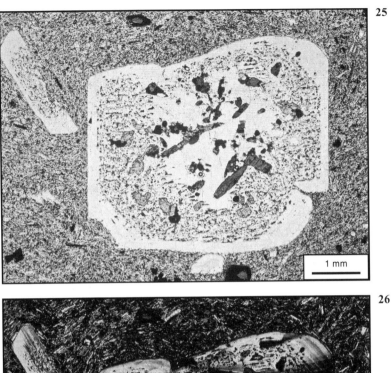

25

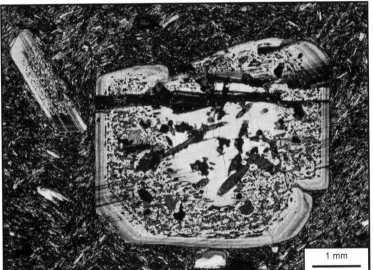

26

25 & 26 *Zoning.* These figures show a phenocryst of plagioclase feldspar in a lava. The innermost zone, commonly referred to as the core, encloses small crystals of other minerals. This is surrounded by a second zone (or mantle) in which there is a high concentration of very small inclusions. Finally the outermost zone (or rim) shows numerous sub-zones or banding, because some bands are nearer their extinction position than others. Notice the zoning caused by the difference in extinction angle can only be seen in the view with crossed polars (**26**).

21

Zoning is the term used to describe changes in a crystal between its core and its outer rim. It may be observed in a number of ways, for example a change in the birefringence (**33**), a change in the extinction angle (**26**) or a change in the absorption colour between the inner and outer parts of the crystal. It usually indicates a change in the composition of the crystal, recording the fact that the fluid from which the crystal grew was also changing composition. Most minerals do not have a fixed chemical composition but belong to a *solid-solution series* and as a crystal grows the outer layer differs in composition from the layer on which it is deposited; this normally results in a change in optical properties which can be detected unless the differences are very slight.

Undulose extinction and sub-grain structure

Some intragranular textures are produced as a result of solid state deformation of the mineral. Under crustal conditions, it is common to observe undulose extinction in quartz grains, due to strain in the mineral lattice, when the angle of extinction changes across a single crystal as the stage is rotated (e.g. 56, 137). Subgrains produce patchy extinction with distinctly bounded areas within a grain with slightly different orientations, but where the misorientation is not high enough (<10°) to form distinct grain boundaries (27). With increasing defomation this process can lead to increase in misorientation until the boundaries can be regarded as grain boundaries and the mineral has recrystallised to form a finer grained aggregate.

Both these features are commonly observed in quartz where they form due to deformation, or as a result of stresses due to cooling and uplift in igneous rocks. Such textures are often preserved in clasts of quartz in sedimentary rocks. In quartz their presence can provide a method for distinguishing quartz from other low relief colourless, low birefringence minerals such as feldspar and cordierite.

This type of deformation is also found in other silicate minerals but generally at high temperatures.

1 mm

27 Subgrains in quartz in crossed polars.

Alteration

A feature which is common to many minerals is *alteration*. Many of the common rock-forming minerals crystallize at relatively high temperatures but when they cool they may be partly replaced by other minerals which are stable at lower temperatures. This is usually associated with to the introduction of water and many alteration minerals are hydrous. Alteration of the primary minerals may take place at any time. The alteration products are commonly too fine grained to be identified optically. However, the observation that some grains are altered and others are not may be of diagnostic importance. For example, in sediments containing quartz and feldspar the latter may often be distinguished because of its alteration (**136, 137**).

In some cases the original shape of the reactant mineral is preserved, producing *pseudomorphs*, e.g., in 28 serpentine is replacing olivine. In 29 fine grained white mica with bright interference colours is replacing the original feldspar which was intergrown with the pyroxene, preserving the granoblastic texture of the high grade rock.

28

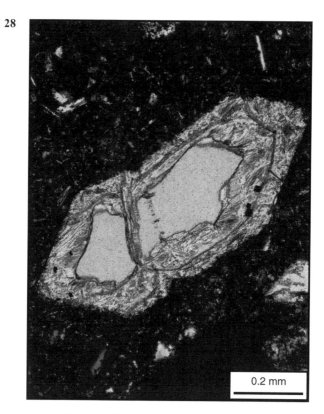

0.2 mm

28   Olivine partly replaced by serpentine which is pseudomorphing the shape of the olivine crystal (crossed polars).

24

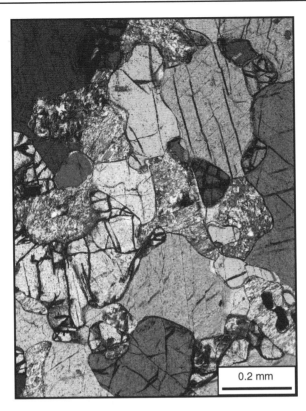

29 Original granoblastic plagioclase feldspar (intergrown with pyroxene) is replaced by a fine grained intergrowth of white mica (crossed polars).

Grain size

Whilst not an optical property of the mineral, the grain size(s) of the minerals is crucial to the classification of most rocks.

> very coarse grained
> 16 mm
> coarse grained
> 4 mm (5 mm often used for igneous rocks)
> medium grained
> 1 mm
> fine grained
> 0.1 mm
> very fine grained
> 0.01 mm
> ultrafine grained

In igneous rocks, grain sizes is a key aspect of classification into volcanic and plutonic rocks (see section 3). In metamorphic rocks, some minerals tend to form larger grains, termed porphyroblasts, set in a finer grained matrix, and this can be helpful in identifying the mineral. For example, while quartz and calcite commonly form the matrix of a metamorphic rock, minerals such as garnet, staurolite, kyanite and andalusite tend to form larger, often euhedral porphyroblasts (e.g. 1, 4).

PART 2

Minerals

One of the first things a student of geology has to learn is the difference between rocks and minerals. Minerals are naturally occurring inorganic chemical compounds with known crystal structures. All rocks, with the exception of those composed mainly of glass, are assemblages of minerals. In many cases, if the crystals are large enough and are coloured differently, we can identify some of the constituent minerals in a hand specimen with the aid of a hand lens. Thus, in a granite we can usually see one, or sometimes two feldspars, a dark mica, and quartz. Before we can identify a rock or begin to describe it, we have to know of what minerals it is composed and to this end we describe some of the common minerals in this section.

Although we have defined a mineral as a chemical compound, the word compound is used in a somewhat different sense from that in which a chemist would use the word. To a chemist, a compound usually has a fixed composition which can be represented by a chemical formula. Common minerals, on the other hand, with some exceptions, are rarely of a single composition. A few minerals are virtually pure compounds, e.g. quartz is almost pure SiO_2; kyanite, andalusite and sillimanite all have the formula Al_2SiO_5 and again only have minor amounts of other elements. Silicate minerals commonly show the greatest complexity in chemical composition and almost all of them are *solid solutions*, i.e., certain elements can substitute for one another in the structure. Thus in the minerals which we call ferromagnesian minerals, magnesium and iron are interchangeable, in the sense that either element may occupy certain sites in the crystal lattice, and in the alkali feldspars sodium and potassium are interchangeable. One of the common minerals, hornblende, embraces a range of chemical compositions in the amphibole group of minerals which represents an even wider range of substitution of different elements in what is essentially one crystal structure.

In this section we have illustrated only a few minerals which are very common and which are necessary for the identification of the majority of igneous and sedimentary rocks. There are a number of minerals which are found only in metamorphic rocks and some of the more common are illustrated in the section on metamorphic rocks. Formulae are given for the common minerals. Some have been simplified to indicate only the chief chemical substitutions, enclosed in brackets. Thus in the case of olivine the composition may lie anywhere between the pure Mg end-member Mg_2SiO_4, and the Fe_2SiO_4 end-member.

A table listing the main optical properties of the key minerals as an aid to identification is included at the back of this book.

Olivine—$(Mg,Fe)_2 SiO_4$

Olivine is the name given to the solid solution series between forsterite (Mg_2SiO_4) and fayalite (Fe_2SiO_4). It is recognized in thin section by its high relief and high birefringence and the fact that it very rarely shows a good cleavage but is commonly traversed by randomly orientated cracks (often containing serpentine formed from the low temperature hydration of the olivine). **30** and **31** show a peridotite composed almost entirely of olivine. **32 & 33** show phenocrysts of olivine in a fine-grained groundmass containing pale brown pyroxene crystals and small lath-shaped plagioclase feldspars with grey or white interference colours. The individual crystals of olivine show different interference colours because they represent different orientations of cutting of the crystals, so they may show first-, second-, or third-order colours. In **33** zoning of the larger olivine crystals is shown by the difference in the interference colours between the main part of the crystals and the rims—the rims are slightly different chemically, being richer in iron.

Olivine is a common constituent of basic and ultrabasic igneous rocks where its composition is magnesium rich: it is usually accompanied by a clinopyroxene which has a brownish colour, whereas the olivine is almost colourless or slightly greenish in colour compared with the pyroxene (see 34). In metamorphosed dolomitic limestones (where dolomite and quartz were present with calcite in the original carbonate rock) the olivine is commonly almost pure forsterite and is colourless in thin section (235,236).

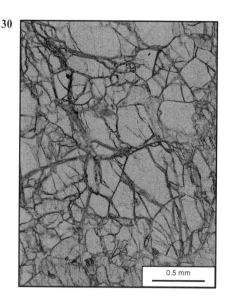

30 Olivine in a peridotite in plane polarized light.

31 Olivine in a peridotite in crossed polars.

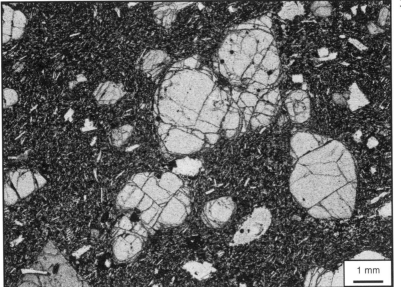

32 Olivine phenocrysts in plane-polarized light.

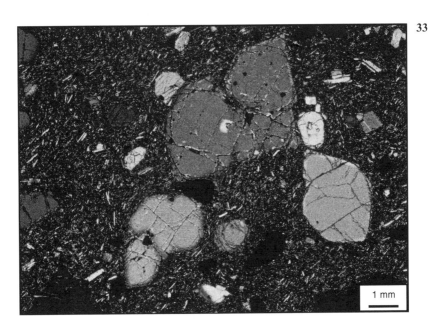

33 Olivine phenocrysts with crossed polars.

Orthopyroxene—(Mg,Fe) SiO₃

The chemistry of this mineral series can be compared with the olivines in that it represents a magnesium-iron silicate series with complete solid solution between the pure magnesium end-member ($MgSiO_3$) and the iron end-member ($FeSiO_3$): the orthopyroxenes however contain more SiO_2 than the olivines.

34 & 35 was taken in plane-polarized light. The coloured crystals are orthopyroxene and the rest of the field in 35 is occupied by plagioclase, quartz, high relief colourless garnet and an opaque mineral. Some of the orthopyroxene crystals have a pink colour whereas others they have a greenish colour. This pleochroism from pink to green is useful as an indicator of the presence of orthopyroxene, but unfortunately it is not always seen. Some of the crystals show cleavages but irregular cracks are also visible. In **36**, taken with crossed polars, the interference colours are first-order only and this illustrates the low birefringence of this mineral. Crystals of orthopyroxene show straight extinction in all sections showing only one cleavage, in contrast to clinopyroxene in which in some orientations, extinction is oblique.

34

1 mm

34 Orthopyroxene forms the majority of this rock showing pink and green pleochroism in plane-polarized light.

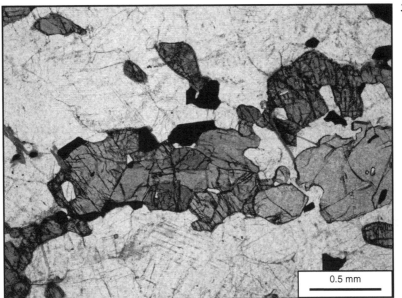

35 Orthopyroxene in plane-polarized light showing different colours of pleochroism – pink to green-depending on orientation.

36 Orthopyroxene taken with crossed polars.

Clinopyroxene—Ca(Mg,Fe) Si$_2$O$_6$

Chemically the commonest clinopyroxenes differ from orthopyroxenes in that the former contain essential calcium. The compositions of clinopyroxenes in basic and intermediate igneous rock are such that they usually lie in the composition range of the mineral known as *augite*.

37 and **38** show large brown coloured phenocrysts of augite in a groundmass of small crystals of augite, olivine and feldspar. In the plane-polarized light view we can see evidence of zoning in the crystal in the bottom right hand quadrant, and in the crossed polars view it is more clearly seen. The two largest crystals can be seen to be composed of simple twins: in some rocks simple twinning is very common in augites. In these photographs the characteristic augite cleavages (but see 10) cannot be seen very well. The birefringence of augite is such that the maximum interference colour is at the top of the second-order. The large twinned crystal at the left edge of the field is showing low first-order colours in both parts because of the orientation in which it has been cut. Clinopyroxenes show oblique extinction of ~40° in sections showing one cleavage, in contrast to clinoamphiboles where the extinction angle is lower, ~20°.

Clinopyroxenes are sometimes green in colour and this may indicate that the mineral and rock are alkali- (Na,K) rich.

37 Clinopyroxene phenocrysts in plane-polarized light.

38 Cinopyroxene phenocrysts taken with crossed polars.

Two-pyroxene intergrowth

39 is a view of a thin section of a rock containing plagioclase feldspar and two pyroxenes both of which are made up of intergrowths. In each of the four quadrants of the figure there are crystals which are almost black and within each crystal there are lamellae showing interference colours. The host crystal in each case is an orthopyroxene nearly at extinction and the lamellae are of clinopyroxene. The other crystals in this field showing red and blue interference colours are clinopyroxenes; these also contain lamellae which are of orthopyroxene. These types of intergrowths can be compared with those in the alkali feldspars (see page 50).

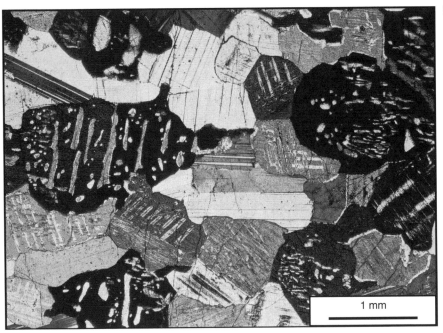

39   Two pyroxene intergrowth with polars crossed.

Amphibole—$NaCa_2(Mg,Fe)_4 \, Al_3Si_6O_{22}(OH,F)_2$

The amphibole group of minerals contains a large number of different solid solutions but all of them have similar crystal structures despite the great variety of chemical substitutions which are possible. There are also quite a variety of colours of amphiboles in thin section and most of them are pleochroic to some extent. The commonest amphiboles in igneous rocks are called hornblendes and the formula given above can be considered to represent a *hornblende*—a general formula for an amphibole is too complex to consider here.

The brown phenocrysts in this view of a thin section of a volcanic rock (**40** and **41**) show pleochroic colours which are different shades of brown—a common colour for hornblendes. Most of the crystals show at least one cleavage and the black rims are due to the formation of iron oxide as a result of oxidation. The angle of 120°/60° between the cleavages in basal sections is very characteristic of amphiboles. The interference colours (**42**) are affected to some extent by the absorption colours but the maximum interference colours of common hornblende are second-order.

Other examples of amphiboles are illustrated in the sections on igneous and metamorphic rocks. Hornblende is also common in metamorphic rocks as well as Actinolite (Fe-rich) –Tremolite (Mg-rich), and also the blue amphiboles, especially Glaucophane (253–255). Ortho amphiboles may also be present which are colourless in plane polarised light. For example, anthophyllite is common in metamorphosed ultrabasic rocks.

40

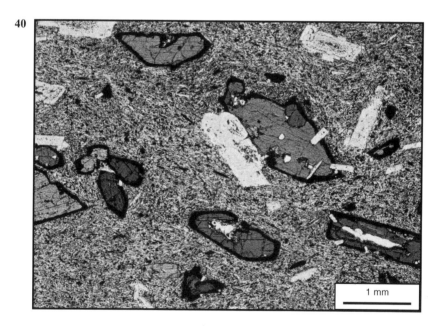

1 mm

40 Amphibole phenocrysts in plane polarized light.

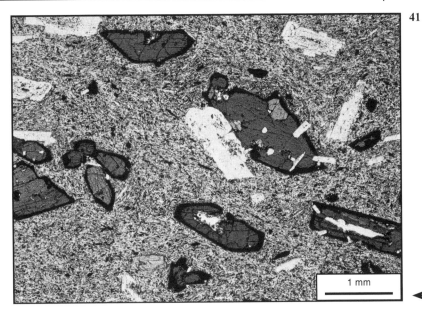

41 Amphibole phenocrysts in plane polarized light. Polars rotated 90° from **40**.

42 Amphibole phenocrysts taken with crossed polars.

Biotite—$K(Mg,Fe)_3AlSi_3O_{10}(OH,F)_2$

Two varieties of mica are common in rocks; colourless muscovite and brown biotite. The mineral with the brown absorption colour in this section is biotite. The above formula shows the usual substitution of Fe for Mg; only the nearly pure Mg end member (phlogopite) has very little colour. Biotite has a perfect cleavage and it is easily split up into thin flexible sheets. In thin section the cleavage can usually be seen and the pleochroism is very obvious in **43** and **44**. The strongest absorption colour is seen when the cleavage is parallel to the polarizer so that in **43** the polarizer was set parallel to the short edge of the figure whereas in **44** it was parallel to the long edge.

The interference colours of biotite are influenced by the strong absorption colour so that we cannot estimate the birefringence easily. Sometimes it is difficult to distinguish biotite from hornblende but when biotite is very close to the extinction position it commonly shows a speckled surface which is quite characteristic. This effect can be seen in one or two crystals in the figure taken with crossed polars (**45**). Biotites are sometimes green in colour, but can be distinguished from green chlorite (**50**) because of the low birefringence of the latter.

43

1 mm

43 Biotite in plane-polarized light.

44

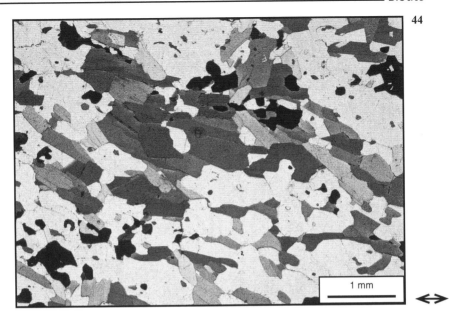

44 Biotite in plane-polarized light. Polars rotated through 90° from **37**.

45

45 Biotite with crossed polars.

Muscovite—$KAl_3Si_3O_{10}(OH,F)_2$

From the formula given above we can see that the chemical difference between muscovite and biotite is that muscovite has no iron and magnesium in its structure and hence it is colourless in hand specimen and thin section. It is described as a 'white mica'. It has a perfect cleavage and this can be seen in some crystals in **46** & **48** taken in plane-polarized light. In **48** & **49** in addition to colourless muscovite, there are a few crystals of biotite and a very high relief mineral which is kyanite (Al_2SiO_5).

The bright interference colours of muscovite are seen in **47** & **49** taken with crossed polars, and it is not readily confused with other minerals. As with biotite, the speckled surface displayed when it is very close to the extinction is quite characteristic. **48** & **49** are an enlarged view of the same thin section used to illustrate kyanite gneiss in the section on metamorphic rocks (**170, 171**).

While muscovite is the most common white mica in metamorphic rocks, other white micas, which are optically almost identical, can be present, such as phengite and pyrophyllite. Hence using the general term 'white mica' can be preferable in some cases.

The only mineral with which white mica can be confused optically is talc, a Mg silicate, which looks almost identical. While much less common, what looks like white mica in an ultramafic rock or marble is more likely to be talc.

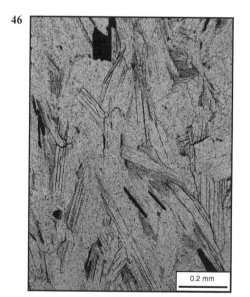

46   Muscovite in plane polarized light.

47   Muscovite in crossed polars.

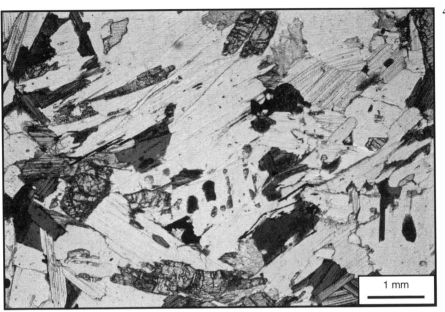

48

48 Muscovite with plane-polarized light.

49

49 Muscovite with crossed polars.

Chlorite—$(Mg, Fe, Al)_6(Si, Al)_4O_{10}(OH)_8$

The green mineral in **50,** taken in plane-polarized light, is chlorite. A green colour is common in chlorite and the reason for its name ('chloros' is Greek for a greenish-yellow colour). In **51** the polarizer has been rotated through 90° and most of the crystals that were green now have a pale straw yellow colour: this pleochroism is characteristic of chlorite. The colourless mineral in **50** and **51** is muscovite. Like the micas, chlorite shows a good cleavage.

The birefringence of chlorite is much less than that of the micas, and chlorites commonly show *anomalous interference colours* (page 16), i.e., colours which do not appear in the interference chart (**19**). The anomalous colours shown by chlorite are usually brown or blue/grey and the former is well seen by the crossed polars view (**52**).

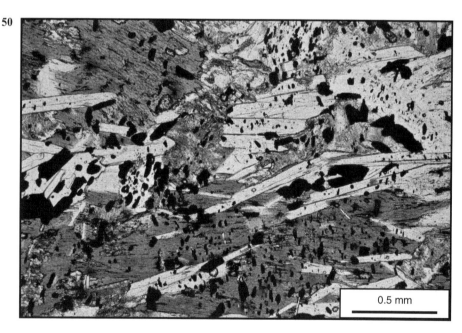

0.5 mm

50 Chlorite in plane-polarized light.

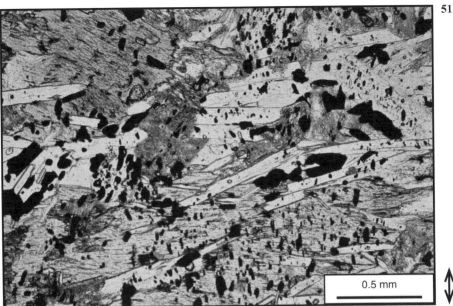

0.5 mm

51 Chlorite in plane-polarized light. Polarizer rotated through 90° from 42.

0.5 mm

52 Chlorite with crossed polars.

Quartz—SiO$_2$

Quartz is one of the most common rock-forming minerals. It is one of the main constituents of granites, sandstones and many metamorphic rocks: its composition is nearly pure SiO$_2$.

It is recognized in thin section by the fact that it is invariably clear and unaltered, it lacks cleavage and with crossed polars shows grey or white interference colours. Because it is so common in rocks it is used along with feldspars to judge the thickness of a thin section – if a yellowish interference colour is seen in quartz this means that the section is slightly too thick. Because it is so ubiquitous we have illustrated it in two different rocks.

In **53** and **54** crystals of quartz and feldspar are seen as large crystals in a fine-grained groundmass. The crystal at the top right part of the field is a badly altered feldspar whereas the clear crystals are quartz. In this rock there are some straight edges to the quartz crystals but there are also embayments suggesting that the growing crystal has incorporated within it, parts of the silicate liquid which later formed the groundmass of the rock.

55 and **56** show a thin section of a granite in which the centre of the field of view is occupied mainly by quartz: we can see, in the plane-polarized light view, that it is clear. Around the edges of the field there are some crystals of biotite and feldspar: alteration of the feldspar can be seen both in the view in plane-polarized light and with crossed polars. A group of quartz crystals near the centre of the field are almost at extinction but the extinction is not uniform. This undulose extinction is a sign that the rock has been strained and is a common feature of much quartz in igneous, sedimentary and metamorphic rocks. Subgrain structures where the grain has started to recrystallize, forming smaller grains with small misorientations across the subgrain boundaries, is also relatively distinctive of quartz in many metamorphic rocks (**27**).

53

54

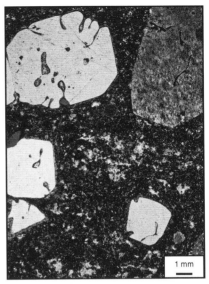

53 Quartz and feldspar crystals in plane-polarized light.

54 Quartz and feldspar crystals with polars crossed.

55

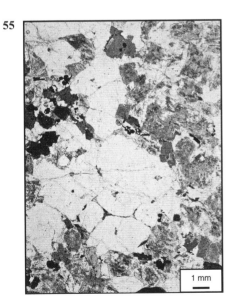

56

55 Quartz in a granite with plane-polarized light.

56 Quartz in a granite with polars crossed.

Feldspars

The feldspars are the commonest of the rock forming minerals in the earth's crust, and comprise two series: the *alkali feldspars* have compositions between $KAlSi_3O_8$ (orthoclase) and $NaAlSi_3O_8$ (albite) and the *plagioclase feldspars* lie between $NaAlSi_3O_8$ (albite) and $CaAl_2Si_2O_8$ (anorthite). Because albite is an end-member of both series the compositions of the feldspars can be represented in a triangle whose corners are these three end members denoted by the abbreviations: Or, Ab and An (**57**).

The plagioclase series is divided into six parts with compositions as follows:

albite = 0–10% An
oligoclase = 10–30% An
andesine = 30–50% An
labradorite = 50–70% An
bytownite = 70–90% An
anorthite = 90–100% An

It is usual to quote the composition of a plagioclase by using the percentage of the end members e.g. $An_{65}Ab_{35}$ or sometimes simply An_{65}. All plagioclase feldspars contain a small amount of K-feldspar, usually less than 5%, and all alkali feldspars contain a small amount of calcium feldspar (<5%) so that in **57** the compositions are shown as a band within the triangle rather than on a line.

In the alkali series there are names for the end members only, since it is difficult to determine the compositions of intermediate members. In addition, intermediate members of the series separate into intergrowths of two feldspars at low temperatures: these intergrowths are known as *perthites* (**61**) or *microperthites*, depending on how coarse the intergrowths are.

The feldspars are important in the most frequently used classification of igneous rocks and it is desirable therefore that we should be able to determine whether two feldspars are present, their relative proportions, and when a plagioclase is present, additionally we should be able to determine its composition by optical methods. In metamorphic rocks the composition of the plagioclase feldspar may indicate the grade of metamorphism.

The material used in preparing thin sections has a refractive index near 1.540. Albite and all other alkali feldspars have refractive indices below this value. Oligoclase has refractive indices near 1.540 but more calcium-rich feldspars have higher refractive indices. Thus if we examine the edge of the slide or any holes in the thin section where a feldspar is adjacent to the mounting material we can, by use of the Becke line (see page 12), determine whether its refractive indices indicate that it is a plagioclase or an alkali feldspar.

All feldspars have relatively low relief and low birefringence so that they are recognized by having grey and white interference colours: only near to anorthite composition does a slight yellowish colour appear in a section of standard thickness. Almost all feldspars have two good cleavages and in some sections they appear to be at right angles to one another. In hand specimens, using a lens, the presence of a cleavage serves to distinguish feldspar from quartz since the latter has no cleavage. Most feldspars exhibit twinning and multiple, polysynthetic or lamellar twinning is very common in plagioclase feldspar: in coarse-grained rocks it can often be seen in hand specimen by using a lens.

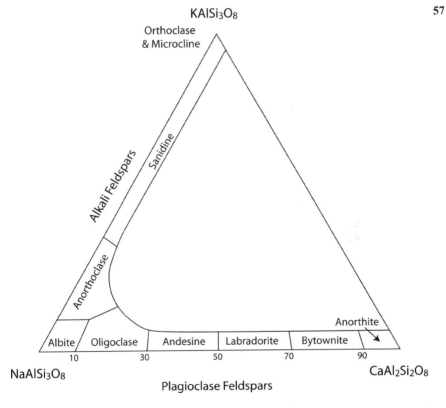

57 Triangular diagram showing the composition of plagioclase and alkali feldspars in terms of the three end members, orthoclase (Or), albite (Ab) and anorthite (An).

Sanidine—(K,Na)AlSi$_3$O$_8$

The potassium-sodium feldspars form a solid solution series at high temperatures but below about 700°C they tend to separate into potassium-rich and sodium-rich parts.

The alkali feldspars in volcanic rocks are commonly known as sanidines and in **58** and **59** some prismatic crystals are shown in a fine-grained groundmass. The preferred orientation suggests that the crystals have been transported by a moving magma before solidification. The crystals show some alteration and a number of them are simply twinned. This habit and the simple twinning are both characteristic of sanidine. It is not easy to determine the composition of an alkali feldspar but potassium rich sanidines are more common than sodium-rich specimens.

Orthoclase is the name given to untwinned or simply twinned potassium-rich feldspar found in many granitic rocks. Since it cannot easily be distinguished from sanidine there is a tendency to restrict usage of the name orthoclase to indicate the potassium end-member of the alkali feldspar series.

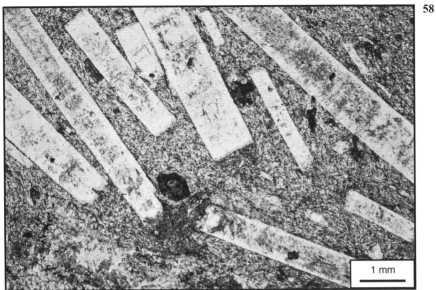

58 Prismatic crystals of sanidine in plane-polarized light.

59 Prismatic crystals of sanidine with polars crossed.

Microcline—KAlSi$_3$O$_8$

Notice that the formula for sanidine involved the substitution of sodium for potassium whereas the formula above indicates that microcline is a potassium mineral with very little sodium present.

60 shows microcline occupying most of the field of view. It is characterized by cross-hatched or "tartan" twinning which is usually enough to indicate the presence of microcline. We have not shown a view in plane-polarized light because it is not useful except to show very low relief. About 10 to 25mm below the top edge of the field there is an intergrowth which consists of quartz and plagioclase. This is known as *myrmekite*.

61 shows a microcline perthite. The vein-like areas are of albite (NaAlSi$_3$O$_8$) and the cross-hatched parts are microcline (KAlSi$_3$O$_8$). This was probably formed as a solid solution and subsequently the two minerals separated to form this perthitic intergrowth.

1 mm

60 Microcline with polars crossed.

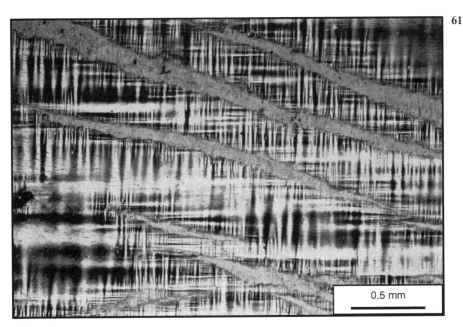

0.5 mm

61 Microcline perthite with polars crossed.

Plagioclase—$NaAlSi_3O_8$-$CaAl_2Si_2O_8$

Plagioclase feldspars invariably have multiple (polysynthetic) twinning which appears as dark and light bands in crystals observed with crossed polars. The most common type of twinning is called *albite twinning* and in this case the twin lamellae lie parallel to a very good cleavage. We have illustrated a twinned plagioclase by a series of three photographs (**62–64**) in which one good cleavage is parallel to the long edge of **62**: the polars are crossed and lie parallel to the edges of the figures. Rotation of the microscope stage in one direction results in one set of lamellae going dark and reaching the extinction position; rotation in the other direction results in the alternate lamellae reaching their extinction position. Twinning according to what is termed the 'albite law' results in the lamellae being mirror images of the adjacent lamellae (without necessarily being the same width) and so the extinction angle in one direction should be exactly equal to that in the other direction if the crystal is cut exactly perpendicular to the cleavage. If the section is not exactly perpendicular to the cleavage, the angles of extinction will not differ by more than a few degrees and the mean value is taken. This is known as the *extinction angle in the symmetric zone.*

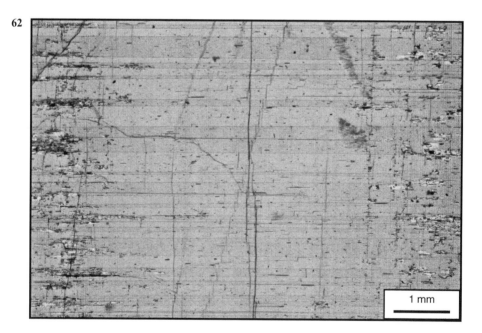

1 mm

62 Twinned plagioclase taken in plane-polarized light.

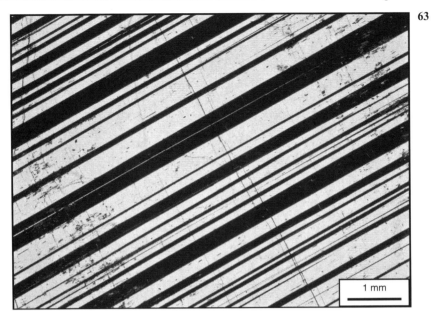

63 Twinned plagioclase taken with crossed polars.

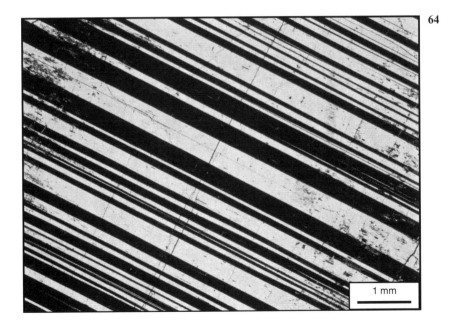

64 Twinned plagioclase taken with crossed polars.

There are two ways in which the measurement of the extinction angle in the symmetric zone may be used:

- The simplest method depends on finding a crystal which is cut not only at right angles to one cleavage but also at right angles to the other perfect cleavage. This cleavage can be seen sub-parallel to the short edge of the figure (**62**) where it appears as a zig-zag line because the two cleavages are not at right angles but are at about 92° to each other. In this case the extinction angles in the photographs (**63, 64**) are 28° and the composition of this crystal can be obtained from the diagram relating extinction angle in sections cut perpendicular to the x crystallographic axis (**66**). In this case the plagioclase is approximately An_{55}—labradorite.
- The second method is used for the more general case in which we cannot find a section orientated perpendicular to both cleavages; we have to use the *maximum* extinction angle in the symmetric zone. It is usual to measure the extinction angle in at least six crystals and to obtain the feldspar composition from a curve relating chemical composition to the maximum extinction angle obtained (**66**).

65 shows part of a large plagioclase phenocryst in a lava: it consists of two parts related by simple twinning. Each of the two parts has multiple twinning. The upper part shows zoning illustrated by the difference in shade of grey interference colour, indicating a different extinction position for parts of the crystal. The reason why the zoning is not shown in the lower half of the crystal is not because it is not present, but because the different orientation of the lower half may not show different shades of interference colour to the same extent as the upper part.

65
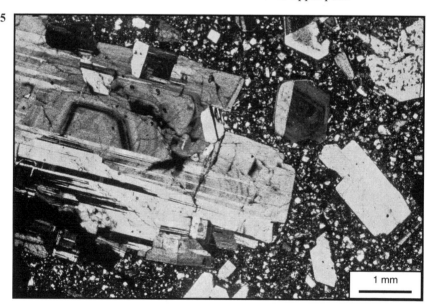

65   Zoned plagioclase phenocryst in a lava with crossed polars.

In low grade metamorphic rocks which are commonly fine grained, albite often does not show evidence of twinning and can be difficult to distinguish from quartz. (240–241) shows a greenschist facies metabasic rock which contains albite as well as quartz.

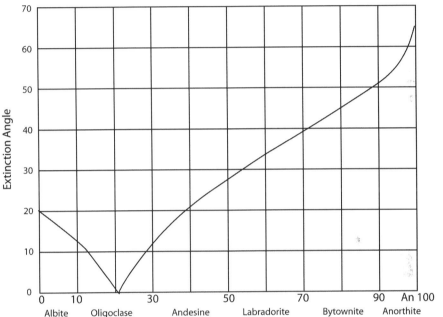

66

66 Diagram showing the relationship between chemical composition and the maximum extinction angle of Albite twins (in plagioclase feldspars) in sections cut at right angles to (010), the composition plane of Albite twins (see text for method). At the sodium-rich end of the series there are two possible compositions for a given extinction angle and they must be distinguished by a refractive index determination. Only plagioclases containing more than 20% anorthite have refractive indices greater than that of the mounting material normally used (c. 1.54).

Craig, J.R. & Vaughan, D.J. 1994. Ore Microscopy and Ore Petrography. Wiley.

Fettes, D. & Desmons, J. 2007. Metamorphic Rocks: A Classification and Glossary of Terms. Cambridge. https://www.bgs.ac.uk/scmr/products.html

Kato, T. 2001. A method to synthesize interference colour chart with personal computer. Jour. Geol. Soc. Japan, 107, 64–67.

Le Bas, M. J. & Streckeisen, A. L. 1991. The IUGS systematics of igneous rocks. Journal of the Geological Society, London, 148, 825–833.

Nepheline—NaAlSiO$_4$

Nepheline is a *feldspathoid* mineral. The feldspathoid minerals are similar in chemistry to the feldspars but have less silica. The formula given above is an ideal formula because all natural nephelines contain some potassium. Nepheline is the commonest of the feldspathoids and its occurrence is an indication that the rock in which it occurs is alkali-rich.

In **67** and **68** phenocrysts of nepheline are shown in a fine-grained groundmass. The crystals are rectangular or sometimes hexagonal in outline. The crystal in the lower left of the field is a broken part of an hexagonal crystal and the crystal at the top edge to the right of centre is also hexagonal. Both of these crystals are nearly black in the crossed polars view because of the direction in which they have been cut.

Nepheline sometimes occurs along with sanidine and it may be difficult to identify individual crystals. If crystals show simple twins they can be readily identified as sanidine. In addition feldspars have better developed cleavages than does nepheline.

Two irregular white areas (plane-polarized light view) in the upper left quadrant are cavities in the rock which appear as holes in the thin section and are thus black in the crossed polars view (**67**).

The minerals nepheline and quartz are not found together in the same rock, but they may be confused because they have similar optical properties. Quartz has a slightly higher birefringence than nepheline and it rarely shows alteration. Quartz does not show rectangular-shaped crystals such as those seen in this section.

67 Nepheline phenocrysts in plane-polarized light.

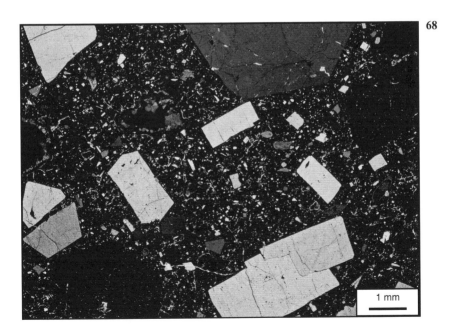

68 Nepheline phenocrysts with crossed polars.

Calcite—CaCO$_3$

The mineral calcite is the main constituent of limestones. Limestones may contain dolomite, CaMg(CO$_3$)$_2$, but this has usually replaced original calcite. Calcite is found in many metamorphic rocks and is the main constituent of marbles; it is also found in some igneous rocks and is the main constituent of a group of rare igneous rocks known as *carbonatites*. A number of methods are used to distinguish dolomite from calcite in hand specimen but here we mention only the simplest chemical test, viz. calcite dissolves with effervescence in cold dilute HCl, whereas with dolomite the reaction is much slower unless the acid is heated.

Carbonate minerals all have very high birefringence so that when viewed with crossed polars they do not usually show interference colours within the range shown in **19** but instead show delicate pastel shades of colour. The high birefringence is the cause of 'twinkling', the name given to a change in relief of the mineral as the microscope stage, or the polarizing filter is rotated. **69** and **70** show part of a thin section in which the polarizer has been rotated through 90°. The effect of this is to change the relative relief of individual crystals so that each crystal looks different in the two figures. Cleavages and multiple twinning are clearly seen. (see also **14** and **15**).

In **71**, taken with crossed polars, the pastel interference colours are seen. One twinned crystal in the lower right quadrant shows some lamellae at extinction since they are nearly black.

69

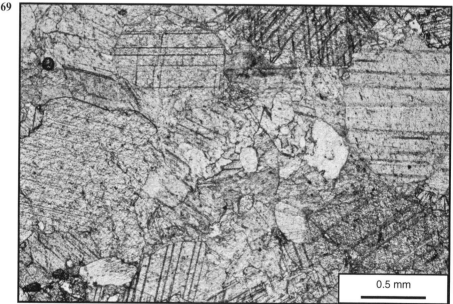

0.5 mm

69   Calcite in plane-polarized light.

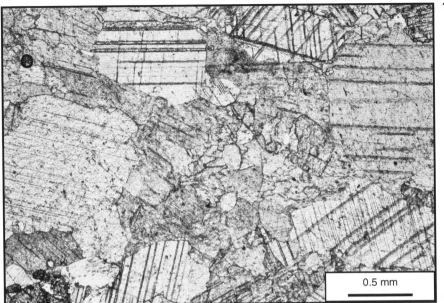

70 Calcite in plane-polarized light; polarizing filter rotated through 90° from **64**.

71 Calcite with crossed polars.

Garnet—$(Fe,Mg)_3Al_2Si_3O_{12}$

The chemical composition given is much simplified but approximates to a garnet of almandine type.

Garnet is common in a great variety of metamorphic rocks. It is frequently red or brownish red in hand specimen, but is normally colourless in thin section although it is sometimes a pale red or brown. Because of its high refractive index and isotropic character, it is generally fairly easily identified in thin section. It tends to form well-shaped crystals although such crystals may be full of inclusions of other minerals.

Some garnet compositions are indicators of formation at rather high pressure whereas other garnets can form at relatively low pressures near to those at the surface of the Earth. There is no simple optical method of determining the composition of garnets – even the refractive index of a garnet is difficult to measure because the liquids of high refractive index used for this purpose are not easily obtainable. The crystals shown in **72** and **73** are euhedral crystals in a metamorphic rock. Their high relief and isotropic character are distinctive.

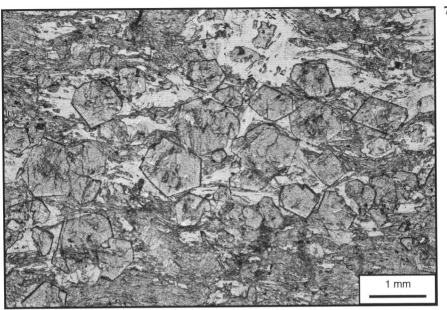

72 Euhedral crystals of garnet in a metamorphic rock, plane-polarized light.

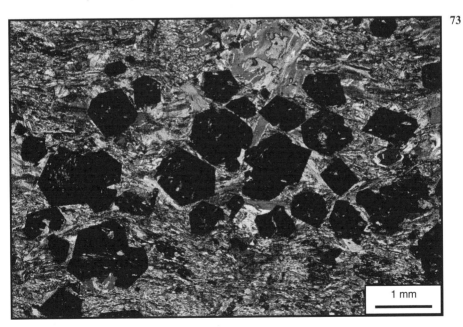

73 Euhedral crystals of garnet in a metamorphic rock, crossed polars.

Opaque minerals

In transmitted light some minerals are not transparent to light even when only 30 μm thick. In optical microscopy they are referred to as the opaque minerals. In some cases the shape of the mineral may help further identification, for example, ilmenite ($Fe^{2+}TiO_3$) tends to occur as elongate grains while magnetite ($Fe^{2+}Fe^{3+}_2O_4$) is cubic and often euhedral. In other instances the rock type may indicate the likely opaque phase.

In order to investigate these minerals further, a reflected light microscope has to be used which delivers light to the highly polished surface of the rock. The light reflected from the surface of the mineral produces a range of optical properties that can be used to identify the mineral. The details of this are beyond the scope of the current book (for details see Craig & Vaughan, 1994) but an example is included here to indicate the usefulness of this technique. **74 & 75** show transmitted light images of a thin section of a peridotite containing abundant opaque minerals as well as plagioclase (low birefringence and showing twinning) and olivine with higher relief and higher birefringence colours. This is a polished thin section and **76** shows the same area but viewed with a reflected light microscope. The silicate minerals are dark grey because they only reflect 5% of the light while the opaque minerals are more reflective and show brighter shades of grey and yellow. The majority of the opaque minerals are Cr-Spinel (light grey, reflecting 12–20% of the light) with brighter sulphide minerals, pyrrhotite and yellow pentlandite (pale yellow), and chalcopyrite (bright yellow). These are often the source of various precious metals, such as the platinum-group elements. Reflected light microscopy is important in the study of metal ore minerals which are often oxides and sulphides and are opaque in transmitted light.

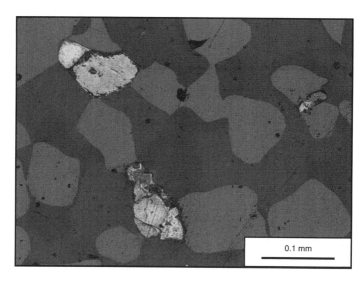

Detail of 76 showing Cr-spinel plus three types of high reflectance sulphide minerals.

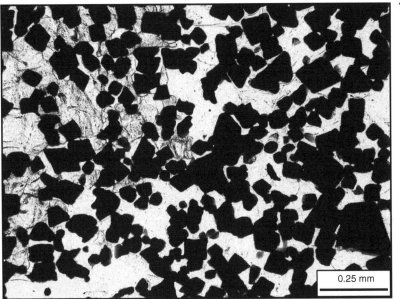

74

74 Transmitted light micrograph showing opaque minerals in plane-polarized light. Gabbro, Rum, N.W. Scotland.

75

75 Transmitted light micrograph showing opaque minerals in crossed polars. Gabbro, Rum, N.W. Scotland.

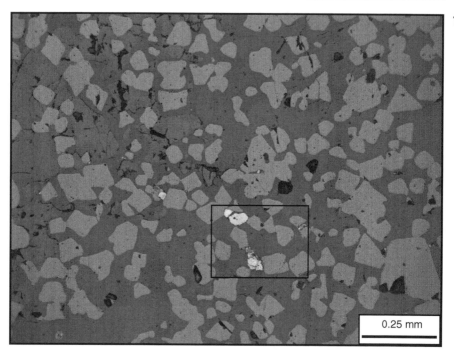

0.25 mm

76 Reflected light photomicrograph of same area as 74 & 75 showing grey euhedral to subhedral chrome spinel, and bright sulphide minerals..

PART 3

Igneous rocks

Igneous rocks are formed by the solidification of magma. Magma is usually a mixture of silicate liquid (melt), crystals of minerals which are crystallising from the liquid and sometimes gas bubbles. If the liquid cools slowly at some depth in the crust the crystals will have time to grow large. Rocks which crystallized in fairly large masses at depths of a few kilometres will form bodies that are termed *plutons* and the rocks are termed *plutonic* rocks. If the magma is erupted from a volcano or from a fissure in the crust it will cool more rapidly and the resulting rock is likely to be composed of very small crystals or glass, but often also containing suspended larger crystals that formed before eruption. Such rocks are described as extrusive or *volcanic* rocks.

A third category of igneous rocks are those which consolidate as dykes or sills and these are termed *minor intrusions* or *hypabyssal* rocks: they are in general of medium grain size.

Another group of rocks which can be regarded as igneous are pyroclastic rocks which are formed of clasts of volcanic material (ash, volcanic bombs and lapilli). These may be directly expelled from volcanoes as clasts or transported and re-deposited via sedimentary processes when they are termed volcaniclastic rocks (see Sedimentary rocks).

We shall see that there are considerable differences in igneous rocks and it is the aim of the petrologist to try to understand what causes this diversity, to determine their relationships to each other and to the geological environments in which they occur. To describe a rock it is desirable to have a system of classification and to give names to each class: we have already made a beginning by considering the circumstances under which the rocks are formed. There are special names for plutonic, hypabyssal and volcanic rocks but the names for the hypabyssal rocks are now rarely used except for the term *dolerite* (in America, *diabase*) for a dyke rock formed from a basaltic magma. Nowadays the plutonic rock name tends to be used for a rock which is coarse grained, i.e., grain size greater than 4 mm, for medium grained rocks (1–4 mm) the prefix 'micro' is attached to the plutonic rock name, e.g. microgranite, and for fine grained rocks (0.1–1 mm) the volcanic rock name is used. Thus the fine-grained chilled margin of a gabbro mass could be described as basaltic. The term aphanitic is used if the crystals are too small to be seen by the naked eye.

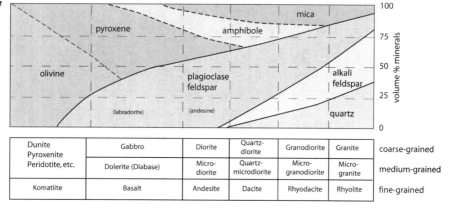

77 Classification of igneous rocks based on mineralogy and grain size. The common rock types illustrated in this atlas, are indicated with an asterix.

Various criteria are used to classify igneous rocks:

Minerals present in the rock (the modal mineralogy): The minerals present in a rock and the relative proportion of minerals (mode) is largely dependent on the chemical composition of the magma. Mineralogy and grain size is the basis of most classification. Generally this works well for coarse grained (plutonic) rocks.

In simplified classification, often most useful to students, igneous rocks are separated on their mineralogy, specifically the presence or absence of quartz, the type and relative proportion of feldspar present (alkali feldspar and plagioclase, and composition of plagioclase), and the type of iron or magnesium (ferromagnesium) minerals present, combined with the grain size (**77**).

The official system most widely used to classify igneous rocks is via the QAPF (Quartz-Alkali feldspar-Plagioclase-Feldspathoid (Foid)) diagrams introduced by the IUGS (International Union of Geological Sciences) led by Streckeisen, based on modal mineralogy. **78 a & b** shows the QAPF diagrams used to classify plutonic and volcanic rocks.

Texture of the rock, especially grain size (note these are only guidelines as there are no agreed grain size divisions for igneous rocks)

Phaneritic – crystals coarse enough to be seen by eye
Sub-divided into   Coarse grained >5 mm
                    Medium grained 1–5 mm
                    Fine grained <1 mm diameter
Aphanitic – crystals too small to be seen by eye
Cryptocrystalline – grains too fine grained to be distinguished with a microscope
Glassy – no crystals formed
Porphyritic – bimodal grain size

Chemical composition: the chemical composition usually reflects the composition of the magma, which determines which minerals will crystallise and in what proportions. This cannot normally be determined from hand specimen or thin section examination, but if chemical analysis can be obtained, various classifications exist (e.g. Le Bas IUGS chemical classification of volcanic rocks or TAS (Total-Alkali-Silica)).

Silica saturation of the rock: if the rock is oversaturated with silica then the mineral quartz will be present, greater than 10% quartz by volume being regarded as oversaturated for the purposes of classification. If the rock is undersaturated with silica it will not contain quartz but other minerals which do not occur with quartz such as feldspathoids. Rocks with less than 10% of either quartz or a feldspathoid are regarded as silica saturated: they cannot contain both quartz and a feldspathoid. This terminology cannot be applied to very fine grained or glassy rocks.

Colour: Terms which are commonly used to describe the colour of rocks are light-, medium-, and dark-coloured, or felsic (usually abundant light coloured feldspar and quartz), mafic (abundant dark coloured ferromagnesium minerals) and ultramafic (almost entirely dark ferromagnesium minerals.) Note that colour can be misleading as glassy rocks with the same composition as a quartz – feldspar rich granite can be black. These terms can be roughly equated with terms which refer to the SiO_2 content of the rock in cases where a chemical analysis of the rock is available. Thus we have the terms *acid, intermediate, basic* and *ultrabasic*. An acid rock is one in which the SiO_2 content is more than 66%; intermediate rocks are 52–66% SiO_2; a basic rock contains 45–52% SiO_2; and for ultrabasic rocks, less than 45% SiO_2. Acid rocks are usually light in colour and basic and ultrabasic rocks are usually dark-coloured. One other term in common usage to discuss the chemistry of rocks is *alkaline*. This term has been very loosely used and we shall not try to define it rigorously here. It is generally used to mean that a rock is richer in alkalis (Na,K) than more common rocks with similar SiO_2 content.

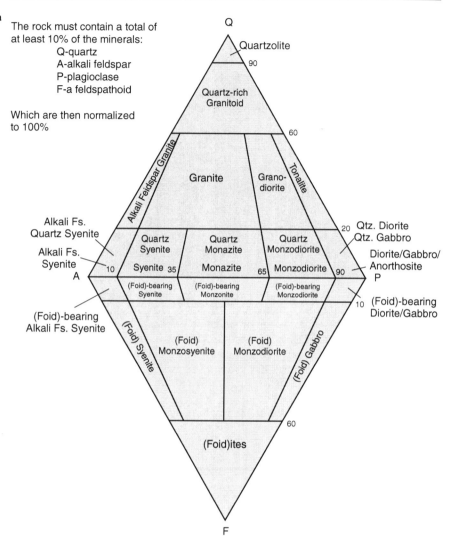

The rock must contain a total of at least 10% of the minerals:
Q-quartz
A-alkali feldspar
P-plagioclase
F-a feldspathoid

Which are then normalized to 100%

Q

Quartzolite

90

Quartz-rich Granitoid

60

Granite

Grano-diorite

Tonalite

Alkali Feldspar Granite

Alkali Fs. Quartz Syenite

Alkali Fs. Syenite

A

10

Quartz Syenite

Quartz Monazite

Quartz Monzodiorite

20 Qtz. Diorite
Qtz. Gabbro

Diorite/Gabbro/
Anorthosite

Syenite 35

Monazite

65 Monzodiorite

90

P

(Foid)-bearing Syenite

(Foid)-bearing Monzonite

(Foid)-bearing Monzodiorite

10 (Foid)-bearing Diorite/Gabbro

(Foid)-bearing Alkali Fs. Syenite

(Foid) Syenite

(Foid) Monzosyenite

(Foid) Monzodiorite

(Foid) Gabbro

60

(Foid)ites

F

78 QAPF classification of igneous rocks (IUGS Subcomission on the Systematics of Igneous rocks (Streckeisen 1974: Le Maitre 2002) for a) plutonic (phaneritic) and b) volcanic rocks, based on the modal mineralogy. This classification is not used when mafic minerals make up >90% of the rock (e.g. peridotites and pyroxenites). The common rock types, illustrated in this atlas, are indicated by an asterisk. When mafic minerals are >90% this classification is not appropriate (see 79).

78b

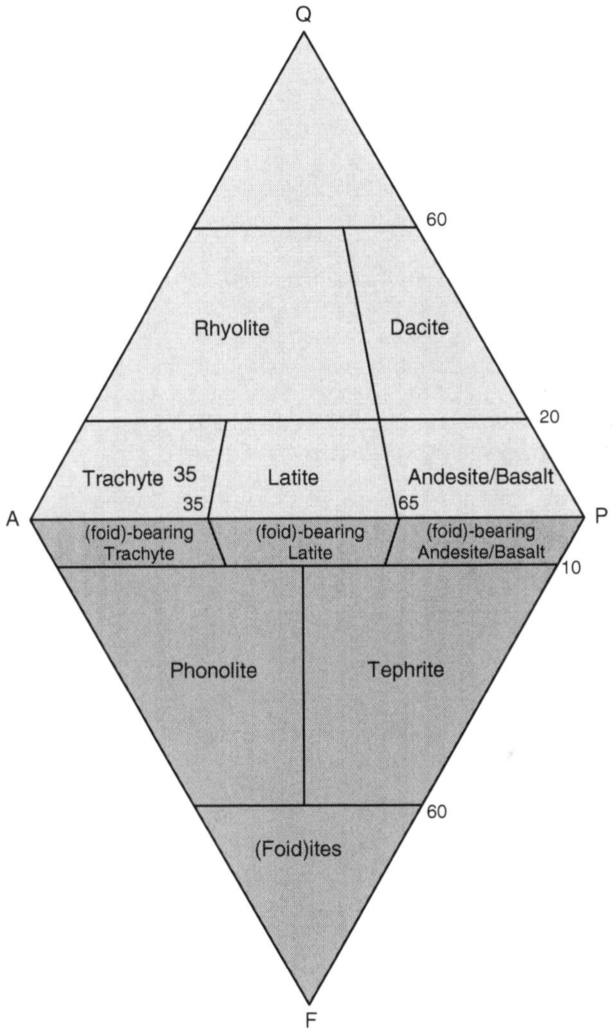
Q
60
Rhyolite
Dacite
20
Trachyte 35
Latite
Andesite/Basalt
35
65
A
P
(foid)-bearing Trachyte
(foid)-bearing Latite
(foid)-bearing Andesite/Basalt
10
Phonolite
Tephrite
60
(Foid)ites
F

The classification based on mineralogy is the simplest method for relatively coarse-grained rocks in which the minerals can be identified first of all by study of a hand specimen with the naked eye or a hand lens and, more precisely, from a thin section examined under the microscope (e.g. **77**). For very fine grained or glassy rocks a classification based on chemical composition will be the most precise method if the facilities are available. If the rock contains >90% of mafic minerals (in peridotites, for example), the QAPF classification (**78**) is also not appropriate, and other classifications based on modal mineralogy are available (e.g. **79** for ultramafic rocks).

A number of igneous rocks are not easily classified by the scheme described above. An example are those rocks that can be grouped under the term *lamprophyre*. This group of rocks occurs as dykes and are characterized by having phenocrysts of a ferromagnesian mineral, biotite, hornblende and less abundant augite, in a groundmass frequently containing a feldspar, which is commonly highly altered, or, in some cases, a feldspathoid. Calcite is a common mineral in the groundmass and is thought to be the result of weathering.

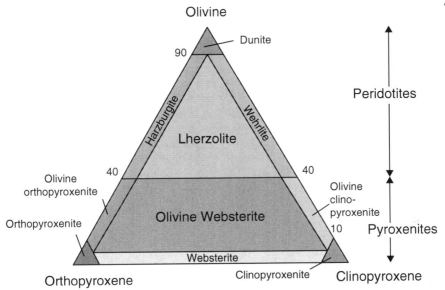

79 Classification of phaneritic igneous rocks containing >90% if ferromagnesium minerals (IUGS Subcomission on the Systematics of Igneous Rocks).

Peridotite

80 is a peridotite consisting of olivine and clinopyroxene and can be classified (see **79**) as a lherzolite (as sufficient orthopyroxene is also present in areas of the rock). All the minerals show relatively high relief and little colour so it can be difficult to distinguish the pyroxenes. The olivine grains have higher relief in plane polarized light (**80**) and also have brighter interference colours in crossed polars (**81**). They show the fractures which are characteristic of olivine. The grain boundaries are very distinct due to minor alteration, probably to serpentine. **82 & 83** shows another peridotite containing olivine with some orthopyroxene, The olivine grains here are small and form a texturally equilibrated granoblastic texture. The orthopyroxene shows cleavage and first order grey interference colours (**83**) and is elongate (in right hand part of view), probably as a result of high temperature deformation. An opaque mineral is also present which is likely to be a Cr spinel.

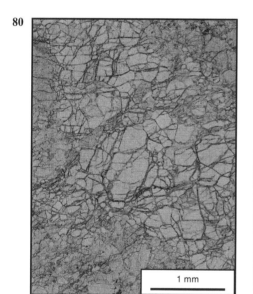

80 Peridotite (lherzolite) in plane-polarized light.

81 Peridotite (lherzolite) with crossed polars.

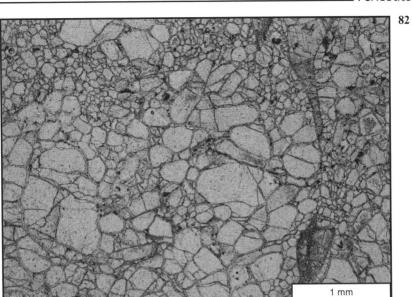

82   Peridotite (Lherzolite) in plane-polarized light.

83   Peridotite (Lherzolite) with crossed polars.

Basalt

This is a fine grained rock with microphenocrysts of euhedral lath-shaped plagioclase, clinopyroxene and olivine, in a fine grained groundmass of the same minerals plus an opaque mineral, magnetite (Fe_3O_4) (**85** & **86**). **84** shows the same rock at higher magnification. The olivine shows the bright blue interference colours in crossed polars (**86**) and irregular fractures, while the clinopyroxene has red-brown colours and a more tabular shape. The relief of the plagioclase (**84** & **85**) is relatively high so it is a calcium-rich composition.

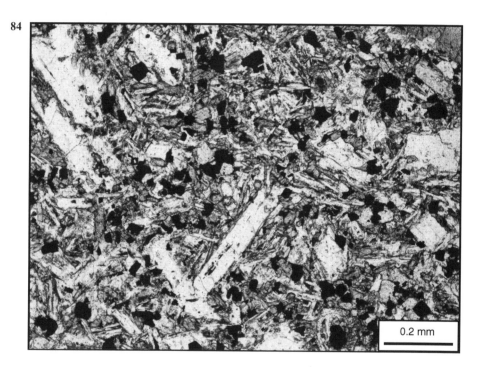

0.2 mm

84 Basalt in plane-polarized light. Locality: Madeira.

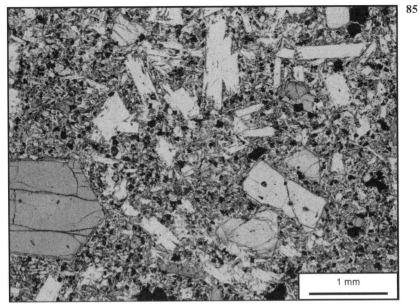

1 mm

85 Basalt, plane-polarized light. Locality: Madeira.

1 mm

86 Basalt with crossed polars. Locality: Madeira.

Olivine basalt

This rock is basaltic but has a much higher proportion of olivine phenocrysts than the previous example shown. The large and medium sized irregularly shaped crystals in **87** and **88** are olivine and these are enclosed in a groundmass of plagioclase feldspar, clinopyroxene and olivine and an opaque oxide mineral. There is a considerable range in size of the olivines which exhibit characteristic cracks and irregular fractures. Basalts with a very high olivine content may be called *Picrites* (which have a high MgO content that is reflected in a high olivine content.)

The different orientations in which the olivine crystals are cut is shown by the range in interference colours from the white colour of the crystal to the left of centre of the field of view to the yellow second-order colour of the crystal to its right (**88**).

Notice that the small lath shaped crystals of plagioclase in some parts of the field of view show a preferential orientation (near vertical in this field of view) which may have resulted from flow of the magma before solidification.

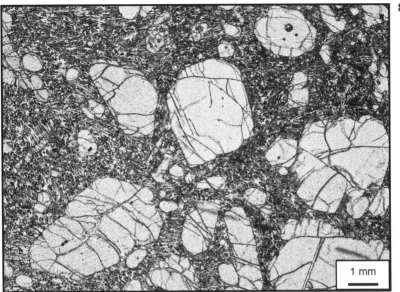

87 Olivine-rich basalt in plane-polarized light. Locality: Ubekendt Island, Greenland.

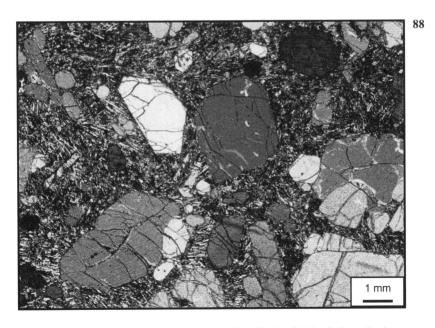

88 Olivine-rich basalt with crossed polars. Locality: Ubekendt Island, Greenland.

Basalt

This basalt (**89**) has a dark coloured glassy groundmass and contains vesicles (round bubbles which are clear in plane polarised light but black in crossed polars, reflecting the isotropic properties of the glass on which the thin section is mounted). Euhedral microphenocrysts of plagioclase (lath shaped crystals showing twinning in cross polarized light), clinopyroxene (pink interference colours) and olivine showing irregular fractures, higher relief and bright interference colours (apart from the grain in the centre which is showing lower brown colours due to its orientation.)

89 Basalt, plane-polarized light. Locality: Stromboli, Italy.

90 Basalt, crossed polars. Locality: Stromboli, Italy.

Dolerite

The rock is a hypabyssal rock of basaltic composition but of medium grain size, consisting mainly of olivine, pyroxene and feldspar, and an opaque mineral. In North America the term diabase is used in preference to dolerite. In **91** & **92** the large crystal in the centre of the field of view (note it all shows blue interference colours) is a clinopyroxene enclosing lath-shaped crystals of plagioclase feldspar. This is characteristic of this type of rock and is known as *ophitic texture*.

93 & **94** show another dolerite with laths of plagioclase, pale brown clinopyroxen and opaque minerals. The plagioclase laths show twinning and some alteration, most obvious in plane polarised light (**93**), making them look cloudy. Olivine is present in the centre of the field of view showing characteristic higher relief, lack of cleavage and high interference colours. Olivine is also present in the top right hand corner of **91** but only showing first order interference colours due to the orientation of the cut.

91

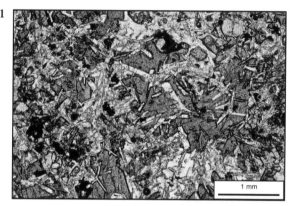

91   Dolerite, plane-polarized light. Locality: Shiant Isles, Scotland.

92

92   Dolerite, crossed polars. Locality: Shiant Isles, Scotland.

80

93   Dolerite, plane-polarized light.

94   Dolerite, crossed polars.

Olivine gabbro

The essential constituents of a gabbro are clinopyroxene and plagioclase feldspar (the latter of labradorite to anorthitic composition). The rock in **95** to **98** has olivine in addition to clinopyroxene and plagioclase, and an opaque mineral. The pyroxenes have a brown absorption colour in the plane-polarized light view (**95 & 97**) and most of the crystals show one or two cleavages. The olivines have less colour and are traversed by irregular cracks. The olivines have higher interference colours than

95

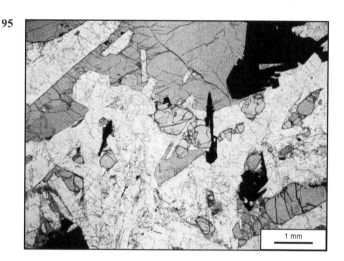

95 Gabbro, plane-polarized light. Locality: Madeira.

96

96 Gabbro, crossed-polars. Locality: Madeira.

the pyroxenes. There is a distinct preferred orientation of the plagioclase crystals in **97 & 98**; potential causes of this fabric could be flow of the crystals in magma or gravitational settling of platy crystals out of the magma.

95 & 96 show a higher magnification view from a gabbro with smaller colourless olivines (e.g., centre) and clinopyroxene, where the plagioclase laths are more random. Opaque minerals are also present.

97

97   Olivine Gabbro in plane-polarized light. Locality: Ardnamurchan, Scotland.

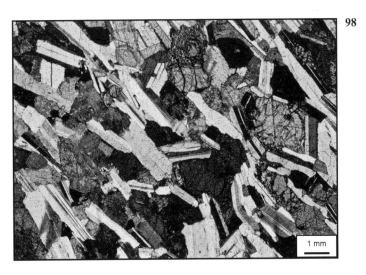

98

98   Olivine Gabbro in crossed polars. Locality: Ardnamurchan, Scotland.

Gabbro

In this gabbro (**99, 100**) there is not as much olivine in the field of view as in the gabbros shown in **95** to **98**, so we have omitted olivine from its name. At the top left corner of the plane-polarized light view (**99**) an olivine crystal can be seen. It has higher relief and is near its extinction position so it appears as nearly black in the field of view with crossed polars (**100**). In the bottom right hand corner is another small olivine crystal. Most of the rest of the field is occupied by twinned plagioclase and clinopyroxene. From extinction angle measurements the plagioclase composition is about $Ab_{30}An_{70}$, i.e., between labradorite and bytownite. The fact that the interference colour is slightly yellow is an indication that either the plagioclase is fairly calcium rich or that the thin section is slightly too thick.

The clinopyroxene has two interesting features which are worth commenting on. Some of the crystals show a slight change in the interference colours at the edges of the crystals due to a change in composition of the pyroxene. Within some of the crystals a fine lamellar structure can just be seen. This is due to exsolution of a calcium poor pyroxene from the calcium rich pyroxene host (see pyroxenes, page 38). A pyroxene crystal left of the centre of the field shows a red-brown colour for one part of a simple twin and yellow for the other part.

99 Gabbro in plane-polarized light. Locality: New Caledonia.

100 Gabbro in crossed polars. Locality: New Caledonia.

Andesite

This rock (**101, 102**) can be seen to consist mainly of microphenocrysts of two minerals in a fine-grained groundmass in which one of the constituents is plagioclase feldspar. The feldspar microphenocrysts are clear in the plane-polarized light view (**101**) and show grey to white interference colours and multiple twinning in the view with crossed polars (**102**). The approximate composition of the plagioclase feldspar can be obtained from extinction angle measurements and is found in this case to be andesine. In addition to multiple twinning, zoning can be seen in some of the feldspar crystals.

The brown-coloured crystals are of an amphibole whose composition cannot be obtained by optical methods but is a hornblende. A few small crystals of a pyroxene are also present: they are colourless in the plane polarised light view and show bright interference colours in the crossed polars view. One such crystal, visible in the middle of the top edge of the figure, shows a red interference colour.

101

101   Andesite in plane-polarized light. Locality: Bolivia.

102

102   Andesite in crossed polars. Locality: Bolivia.

Diorite

We can estimate from the plane-polarized light view of this rock (**103**) that it is made up of about 25% by volume of dark minerals and the other 75 vol.% is mainly plagioclase feldspar. Biotite is fairly easy to identify from the brown colour of the crystals and its good cleavage. There are two pyroxenes in this rock and they can be distinguished by their birefringence because the orthopyroxenes show only first-order grey interference colours whereas the clinopyroxenes show first- and second-order colours (**104**).

Just to the left of the centre of the field of view is an orthopyroxene crystal showing a first-order grey colour and adjacent to it is a clinopyroxene showing red and blue interference colours. There is some quartz which is clear in comparison with the feldspar which shows some cloudy alteration but there are only a few small crystals and it is not easy to identify in the photographs. The feldspars show some zoning as well as twinning and have just slightly higher refractive indices than quartz: they are of andesine composition.

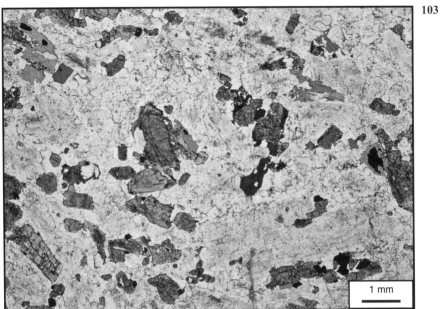

103

103 Diorite in plane-polarized light. Locality: Comrie, Scotland.

104

104 Diorite in crossed polars. Locality: Comrie, Scotland.

Granodiorite

The coloured minerals in **106** are biotite and hornblende. At the top edge of the field we can see a biotite crystal with a deep brown colour in the plane-polarized light view (**106**). Biotite shows pleochroism and other crystals show lighter brown and green colours due to their different orientations. To the right are some paler coloured green crystals which are showing blue interference colours in the crossed polars view (**107**)—these are amphiboles. A large area just below the centre of the field is dark grey in the crossed polars view; this is a highly altered plagioclase. In the plane-polarized light view the areas that are clear are mostly quartz but some alkali feldspar is also fairly clear and unaltered. It is very difficult to estimate the amount of alkali feldspar when it is untwined, as in this case, and it is sometimes necessary to stain the thin section with a chemical which colours the alkali feldspar but not the plagioclase and quartz.

Plagioclase is the dominant feldspar and in the granodiorite shown in **105** it shows concentric zoning. Here the quartz grain to the left shows some subgrains which help distinguish it from feldspar.

105

1 mm

105 Granodiorite in crossed polars.

106 Granodiorite in plane-polarized light. Locality: Moor of Rannoch, Scotland.

107 Granodiorite in crossed polars. Locality: Moor of Rannoch, Scotland.

Rhyolite

This rock (**108, 109**) contains phenocrysts of feldspar in a groundmass which is glassy but is full of tiny laths of alkali feldspar. The concentric cracks in the glassy groundmass are known as *perlitic* cracks.

Most of the phenocrysts contain inclusions of glass and both alkali feldspars and plagioclase are present, but it is difficult to be certain of the identity of each crystal. Simple twinning is usually an indication of sanidine (a high-temperature variety of alkali-feldspar) and lamellar twinning indicates plagioclase. There are no obvious phenocrysts of quartz and because of this we cannot be certain that this is a rhyolite without analysing the rock chemically, which has been done in this case.

In addition to the feldspar phenocrysts there are microphenocrysts of clinopyroxene showing bright interference colours.

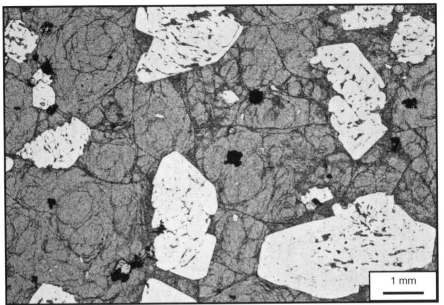

108

108   Rhyolite in plane-polarized light. Locality: Eigg, Scotland.

109

109   Rhyolite in crossed polars. Locality: Eigg, Scotland.

Microgranite

The rock shown in **110** and **111** has a very low proportion of mafic minerals and consists mainly of alkali feldspar and quartz. The alkali feldspar is rather altered and appears brown in the plane-polarized light view (**110**), whereas the quartz is clear. The intergrowth of alkali feldspar and quartz is probably the result of simultaneous eutectic crystallization of the two minerals and this texture is described as a *granophyric* texture.

The ferromagnesian mineral could not be identified from these photographs because it is also rather altered.

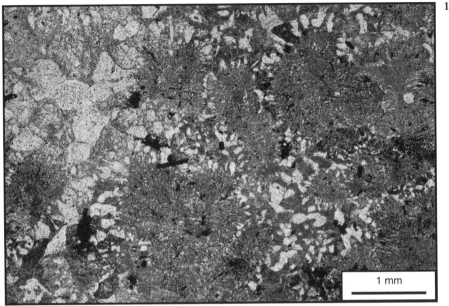

110 Microgranite in plane-polarized light. Locality: Skye.

111 Microgranite in crossed polars. Locality: Skye.

Granite

Two views in plane-polarized light (**112, 113**) show pleochroism of the biotite. There are two feldspars – alkali feldspar and plagioclase, the alkali feldspar being present in a much higher proportion than the plagioclase. The plagioclases are more altered than the alkali feldspar and have slightly higher relief. Most of the alkali feldspar crystals show patches of polysynthetic twinning (see page 52) sometimes only in the cores of the crystals. The clear areas are quartz.

This rock (Westerly Granite) was chosen to be used for tests of silicate analyses because it was relatively unaltered and of uniform grain size so that after crushing a large number of samples of similar composition could be obtained and sent to geochemists throughout the world.

112

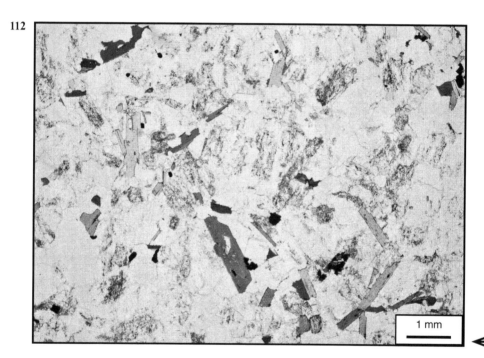

1 mm

112 Granite in plane-polarized light. Locality: Westerly, Rhode Island, USA.

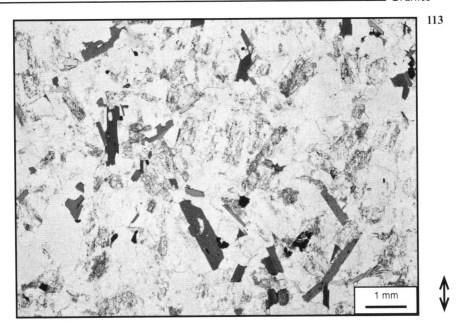

113 Granite in plane-polarized light with polarizing filter rotated through 90° from 112. Locality: Westerly, Rhode Island, USA.

114 Granite in crossed polars. Locality: Westerly, Rhode Island, USA.

Alkali granite

In the plane-polarized light view of this rock (**117**) there is very little relief or alteration of the feldspars to help in distinguishing them from quartz. The coloured crystals in this view are almost all the same mineral, an alkaline amphibole showing very distinct pleochroism from a brown colour to a deep prussian blue. The interference colours, in the view with crossed polars (**118**), are strongly affected by the absorption colour so that we could not estimate the birefringence from the interference colour alone. In **117** the rounded white area in the centre of the left half of the field is quartz as is an adjacent grey-coloured region to the right. Large black areas of similar size are of quartz (along with some irregular opaque minerals). Within the quartz crystals, and filling the regions between them, are small crystals of albite and microcline. Both these minerals have refractive indices below that of quartz and we can distinguish them by their twinning. The microcline shows cross-hatched twinning whereas the albite shows lamellar twinning. Some muscovite mica is also present in the upper edge of the field of view, shown in crossed polars (**118**) by the bright orange interference colours but which is colourless with low relief in plane-polarised light.

115 & **116** show a coarse-grained alkali granite containing interlocking rectangular alkali feldspar, distinguished in plane polarised light from the irregular shaped quartz by the cloudy alteration and in crossed polars by the perthitic texture (exsolution of plagioclase lamellae in dominantly alkali feldspar). The dark green/brown mineral is a sodic pyroxene, aegerine. The interference colours are strongly affected by the absorption colour.

115

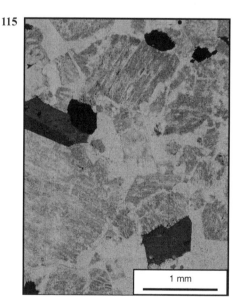

116

115 Alkali granite in plane-polarized light. Kugnat, Greenland.

116 Alkali granite in crossed polars.

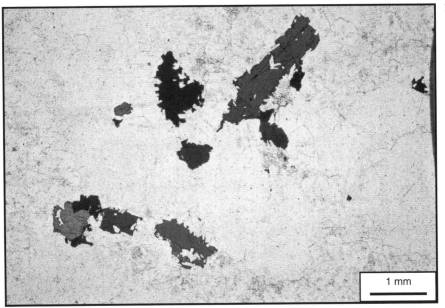

117 Alkali granite in plane-polarized light. Locality: Jos, Nigeria.

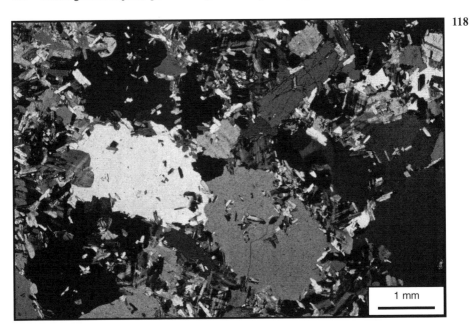

118 Alkali granite in crossed polars. Locality: Jos, Nigeria.

Phonolite

Phonolite is the volcanic equivalent of a nepheline syenite so that its essential constituents are nepheline and alkali feldspar, generally with a small amount of alkali pyroxene. This rock (**119, 120**) contains euhedral phenocrysts of nepheline in a groundmass of nepheline and lath-shaped alkali feldspar crystals. There are also microphenocrysts of nepheline and lath shaped, brownish-green pyroxenes. Some of the nepheline crystals are rectangular in outline and some are hexagonal. Those that are hexagonal are black or very nearly black in the view with crossed polars (**120**). Since the birefringence and refractive indices of nepheline and feldspar are very similar it is difficult to distinguish them but a few differences can be noted: nepheline does not have two perfect cleavages like feldspar, it does not form simple twins and it has straight extinction in all sections whereas alkali feldspars have straight extinction only in some sections.

It is useful to note that in rocks rich in alkalis the pyroxenes are often green in colour in thin section and amphiboles may be bluish green to dark indigo blue.

119

119 Phonolite in plane-polarized light. Locality: Comoro Islands, Indian Ocean.

120

120 (image 95) Phonolite crossed polars. Locality: Comoro Islands, Indian Ocean.

Nepheline syenite

Photographs **121** and **122** show tabular phenocrysts of alkali feldspar which have suffered alteration as indicated by their brownish colour seen in the plane-polarized light view (**121**). The other essential mineral in this rock is nepheline. The ground-mass of the rock is made up of both nepheline and feldspar. The birefringence of nepheline and feldspar are very similar so they may be difficult to distinguish but a number of features permit us to identify the different phenocrysts in this rock. The alkali feldspar has a tabular habit and many of the crystals display simple twinning. This characteristic was probably inherited from a sanidine precursor (see page 52); these crystals are now microperthite, i.e., they have unmixed to a sodium-rich and a potassium-rich feldspar.

The higher magnification view (**123**) was taken to show:

- the clear unaltered nepheline in the bottom left of the figure.
- a few dark minerals, viz. greenish pyroxene on the left and brown biotite to the right.

The proportion of dark minerals is very small as can be seen from the lower magnification view (**121**).

121

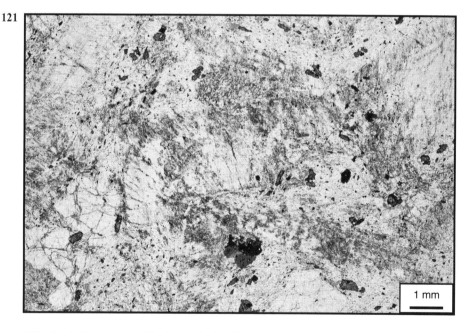

1 mm

121 Nepheline syenite with plane-polarized light. Locality: Barona, Portugal.

122   Nepheline syenite in crossed polars. Locality: Barona, Portugal.

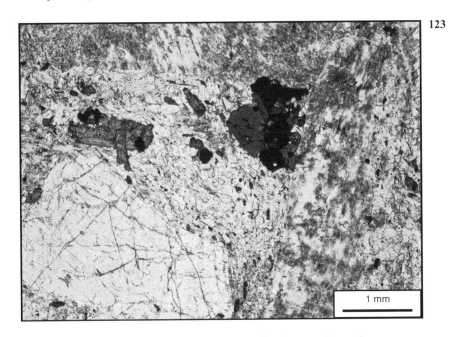

123   Nepheline syenite in plane-polarized light. Locality: Barona, Portugal.

Lamprophyre

Lamprophyre is the name for a group of rocks which occur as dykes and have one of the ferromagnesian minerals as phenocrysts, and a groundmass of alkali feldspar, plagioclase or sometimes a feldspathoid. They are considered to have similar origins as kimberlites. The feldspar is commonly badly altered. Primary minerals are commonly altered in these rocks due to their high volatile content.

1**24** and **125** show a micaceous lamprophyre and the rims of some of the biotite crystals can be seen to be slightly darker than the cores; this may be due to a change in composition possibly caused by partial oxidation of the iron. The groundmass feldspar is badly altered and has a brownish colour slightly paler than the biotite. From its low refractive index it is probably albitic in composition. There is quite a high proportion of calcite in the interstices of this rock and the clear area at the top left (**124**), filled with calcite. The high refractive index acicular crystals with low birefringence are apatite, a calcium phosphate mineral and a common accessory mineral in lamprophyres.

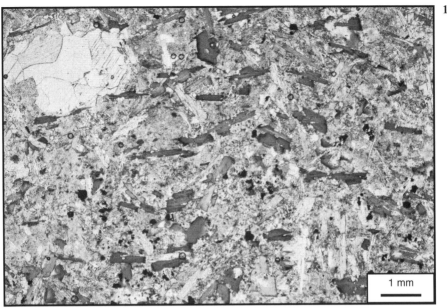

124 Lamprophyre in plane-polarized light. Locality: Ross of Mull, Scotland.

125 Lamprophyre in crossed polars. Locality: Ross of Mull, Scotland.

Ignimbrite

An ignimbrite is a rock formed by the solidification of hot fragments explosively erupted from a volcano and forming a pyroclastic flow. It is composed of volcanic ash, pumice and lithic fragments in a matrix of vitric, crystal and lithic ash. In thin section the textures can be variable due to differences in compaction and welding. If sufficiently hot, the grains may weld together producing a welded ignimbrite containing flattened pumice fragments called fiamme (Italian: flame).

The sample shown here consists of quartz and feldspar crystals embedded in a groundmass or matrix of volcanic glass fragments. The view with crossed polars (**127**) is very black because of the high proportion of glass in the rock. Under the weight of overlying material the hot glass fragments are flattened and welded together . The partial crystallization in the matrix may have occurred during the cooling of the pile of pyroclastic material. Large fragments of pumice have been compacted to produce fiamme, although this sample is not strongly welded.

At the lower edge of the field in the plane-polarized light view (**126**) we can see that the glass shards have become preferentially orientated parallel to each other as a result of flattening after emplacement (due to rheomorphism).

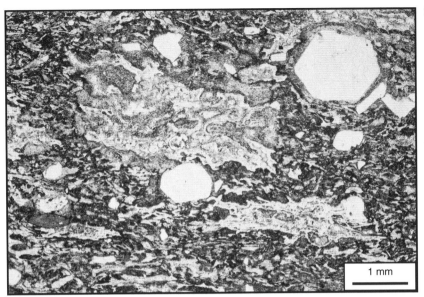

126 Ignimbrite in plane-polarized light. Locality: Taupo Volcanic Zone, North Island, New Zealand.

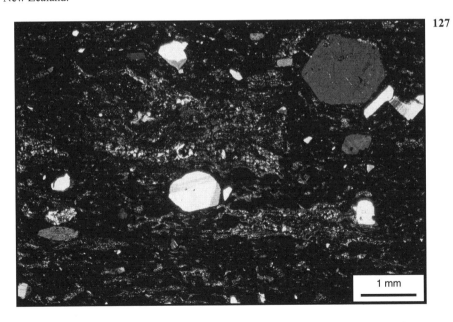

127 Ignimbrite in crossed polars. Locality: Taupo Volcanic Zone, North Island, New Zealand.

PART 4

Sedimentary rocks

Sedimentary rocks are composed of two major groups:

- *Terrigenous clastic rocks*, which are largely composed of fragments of pre-existing rocks and minerals transported from their source in a fluid (air or water) and deposited.
- *Rocks formed by precipitation from solution*, either from the secretions of organisms, as in many limestones, or directly as, for example, in the case of salt deposits.

Sedimentary petrography usually refers to the study of sedimentary rocks under the microscope. It is important since it is often the only easily available method of studying the detailed mineralogy and grain types of sediments. It can reveal the original source of the eroded fragments of terrigenous clastic rocks (*provenance*) and shed light on the depositional environment of limestones. Microscopic studies are particularly important in understanding post-depositional changes which occur in sediments. These changes, known as *diagenesis*, include physical and chemical modifications which occur during burial as a result of increasing load pressure and the passage of fluids through the sediment. Diagenesis may profoundly affect *porosity* (the percentage of pore space in a bulk volume of rock) and *permeability* (ability of a rock to allow fluid to flow through it). This is of great relevance to the study of aquifers and hydrocarbon reservoirs.

Ideally petrological studies of sedimentary rocks should be integrated with field data in the case of outcrop studies, or well-log data in the case of subsurface studies, in order to unravel the depositional and post-depositional history of sedimentary sequences.

Terrigenous clastic rocks

The primary division of terrigenous clastic rocks is according to average grain size. It is dominantly the *sandstones* or *arenites* (average grain size range from $\frac{1}{16}$ to 2 mm) which are studied using the petrological microscope.

In many finer grained sedimentary rocks (mudstones) the particles cannot be easily studied without special techniques or by the use of an electron microscope. In coarser rocks (conglomerates) the grains can usually be identified in hand specimen using a lens. Furthermore, the small area of a typical thin section will contain relatively few grains of a coarse-grained rock, which may not be representative of the rock sample as a whole.

In describing a sandstone it is usual to consider it under the following headings: the grains or particles, matrix, cement and pores.

Grains The most commonly encountered mineral grains are quartz and feldspar. However, many sedimentary rocks contain grains which are recognizable fragments of the source rock and usually contain more than one mineral, and so are called *rock fragments*. These three grain types form the basis of most sandstone classifications (**128**) although other mineral grains such as micas may be present.

Matrix Matrix is the fine-grained sediment between the principal clastic particles. Much of it is clay-sized material too fine to resolve with a light microscope. It normally comprises fine quartz and clay minerals. Clay minerals are a complex series of hydrous alumino-silicates which mostly form from the chemical weathering of silicate minerals from the source rock. Matrix is absent from many 'clean' sandstones. If present in small amounts the sandstone can be said to be 'muddy'. If more than 15% of the total rock volume is matrix the sandstone is called a *wacke* or *greywacke* and classified separately from arenites (**129**). Clay matrix may be deposited at the same time as the sand particles. On the other hand it may have infiltrated after deposition or be formed diagenetically from the breakdown of chemically unstable rock fragments within the sediment.

Cement Cement is the term used to denote the crystalline material precipitated in spaces between the grains. In most sandstones it is quartz or calcite. It is entirely post-depositional and may not form until millions of years have elapsed after deposition of the grains. Cement is the principal lithifying material which converts a loose sediment into a sedimentary rock.

Porosity Any spaces not occupied by grains, matrix or cement are pores. In the subsurface these will normally be occupied by water or more rarely by oil or gas. Pores will lose any water during section making and will be filled with air or with an impregnating medium in thin section. Porosity may be described as *primary* in which case it has been a feature of the rock since formation, or *secondary* where it has developed after deposition. Most primary porosity in sedimentary rocks is intergranular (interparticle) where it is between the grains or intragranular (interparticle) if it lies within a grain such as a fossil fragment. Secondary porosity forms as a result of fracturing or by solution. Solutional porosity may pick out individual components such as shells (mouldic porosity) or create vugs or caverns particularly in limestones.

Classification Several sandstone classifications have been proposed based on the abundances of the components described above. We give an example of classification using QFR diagrams. These are triangular diagrams with quartz, feldspar and rock fragments at the poles. The main triangle is for arenites, sandstones containing less than 15% fine-grained matrix (**128**). The volume of all components is estimated and the quartz, feldspar and rock fragment components recalculated to total 100%. Granite and gneiss fragments, where present, plot with feldspar in

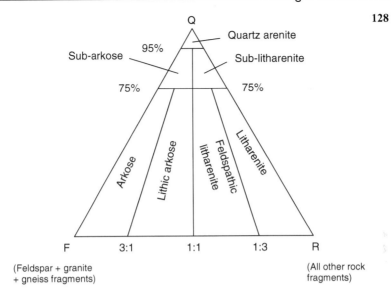

128 Folk classification of sandstones containing less than 15% matrix.

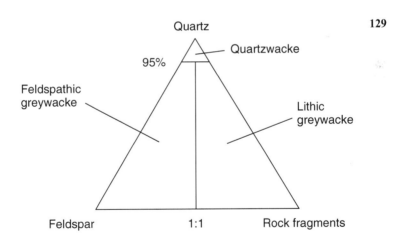

129 Classification of sandstones with more than 15% matrix (greywackes).

this classification rather than with rock fragments. This is because most feldspar derives from weathered granites and gneisses and as far as possible fragments with a common origin should be plotted together. In using this diagram it is best to take the quartz percentage first. Any rock with more than 95% quartz is a *quartz arenite*. A rock with 75–95% quartz is a *sub-arkose* if feldspar is more common than rock fragments and a *sub-litharenite* if rock fragments dominate over feldspar. Rocks with less than 75% quartz are classified according to the ratio of feldspar to rock fragments.

The smaller triangle (**129**) shows the classification of the greywackes (sandstones with more than 15% fine-grained matrix) using similar QFR poles.

Textural features Textural features of sediments such as grain shape and roundness can also be visually estimated from thin sections. *Sorting* of sediments is also important. Sorting indicates the grain size distribution of the sediment sample – a rock with grains all of much the same size is said to be well-sorted whereas a rock with a great range of grain sizes is said to be poorly-sorted.

It must be remembered that in a thin section, grains will not all show their true maximum diameter. Hence even well-sorted sediments rocks show an *apparent* variation in grain size diameter greater than the real variation.

Carbonate rocks

Carbonate rocks are dominantly composed of two minerals – calcite, $CaCO_3$, and dolomite $CaMg(CO_3)_2$. In many recent shallow marine carbonates the mineral aragonite, also $CaCO_3$, is common. However, it is metastable and during diagenesis is likely to dissolve or to recrystallize to calcite. Dolomite is a secondary mineral which replaces calcite or aragonite, or forms a cement. The replacement may take place early in diagenesis soon after deposition, or much later during burial.

The two common carbonate minerals have similar optical properties with variable relief and high-order interference colours and cannot always be easily distinguished by optical microscopy.

The principal constituents of limestones are the organized grains made up of calcium carbonate known as allochemical components, micrite and sparite.

Allochemical components These, often abbreviated to allochems, are organized aggregates of carbonate which have formed within the basin of deposition. They include ooids, bioclasts, peloids and intraclasts.

Ooids These are spherical or ellipsoidal grains up to 2 mm in diameter which have regular concentric laminae of fine-grained carbonate developed around a nucleus. They form by precipitation from supersaturated solution while held in suspension in turbulent waters.

Peloids These are the allochems which are composed of fine-grained carbonate lacking any recognizable internal structure.

Intraclasts These are composed of sediment once deposited on the floor of the basin of deposition which was later eroded and reworked within that basin of deposition to form new grains.

Bioclasts These are the remains, complete or fragmented, of the hard parts of carbonate-secreting organisms.

Micrite This term is short for microcrystalline calcite and refers to carbonate sediment with a crystal size less than 5 μm. It forms in the basin of deposition either as a direct precipitate from seawater or from the disintegration of secretions of calcium carbonate associated with organisms such as algae. *Carbonate mud* is a term which is often used interchangeably with micrite although strictly mud includes material up to 62 μm in size. The crystal size of micrite is much less than the thickness of normal thin sections and so it is not possible to make out individual crystals under the microscope. Micrite often appears medium to dark grey. The outer parts of ooids, peloids and intraclasts are made of micrite.

Sparite The term 'sparite' is short for sparry calcite and refers to crystals of 5 μm or more in diameter. Much of it is a lot coarser with crystals typically tens to hundreds of microns in size. Sparite is a cement (see page 110) and is thus a secondary pore-filling precipitate.

Porosity Porosity is an important component of some carbonate rocks and can be described using the same terminology as that outlined above for terrigenous clastic rocks.

Other components Many limestones are not pure carbonates but contain some terrigenous clastic material. Identifiable grains are usually quartz, and limestones with more than a few percent quartz are called sandy limestones. Limestones, particularly fine-grained limestones, often contain clay and are known as muddy limestones. It is impossible to estimate the percentage of clay mixed with fine carbonate from a thin section. As with sandstones, carbonate sediments may contain pore-space.

Classification Many limestones contain material formed in the depositional area where transport of grains is not a major feature. This means a great range in grain sizes commonly occurs in one sample. Thus grain size classification of limestones is not as significant as it is with terrigenous clastic rocks.

Many detailed limestone classifications have been proposed. The one given here (**130**) is one of the most useful and is based on the depositional texture of the rock. It is useful to modify this classification by adding the name of the dominant allochem type, e.g. a sediment containing ooids cemented by sparite and lacking any car-bonate mud is called an oolitic grainstone and a rock with bioclasts and carbonate mud in which the matrix supported the grains is called a bioclastic wackestone. It is often not easy to distinguish grain from matrix support in a thin section. In general, though, sediments with more than 55–60% grains will be grain-supported even if not every grain is touching another in the section. Remember you are looking at an almost two-dimensional section and grains may be in contact out of the plane of the section.

Original components not organically bound together during deposition				Components organically bound during deposition
Contains carbonate mud			No carbonate mud	
Mud- supported		Grain-supported		
<10% allochems	>10% allochems			
MUDSTONE	WACKESTONE	PACKSTONE	GRAIN-STONE	BOUND-STONE

130 Dunham classification of limestones according to depositional texture. Boundstones are sediments in which the components are organically bound during deposition to form a rigid structure. They include much of the sediment making up reefs and are normally identified at a hand-specimen level rather than microscopically.

Quartz arenite

The rock in **131** and **132** is a quartz arenite, a terrigenous clastic sediment in which the majority of grains are quartz. In this example the quartz grains are single detrital crystals with uniform interference colours. The sediment is poorly-cemented so that substantial pore-space remains, speckled in plane-polarized light (**131**) and black under crossed polars (**132**). The quartz grains have a thin coating containing iron oxide, responsible for the faint orange colour at the edges of some grains. The rock has been moderately compacted and some grains have been welded to their neighbours or penetrate them slightly (e.g. upper left of field of view).

133 shows a medium-grained quartz arenite under crossed polars. This example is lacking significant porosity. Depositional porosity has been reduced by compaction and by precipitation of quartz cement. In this rock it is not possible to distinguish detrital quartz from quartz cement because the cement has grown by extension of the original crystal. Quartz-cemented quartz arenites form very hard durable rocks and are sometimes called **quartzites**.

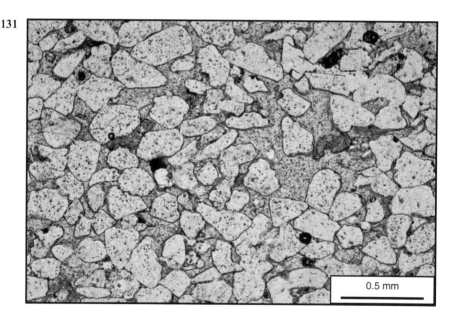

0.5 mm

131   Quartz Arenite in plane-polarized light. Locality: New Red Sandstone, Permo-Trias, England.

132

0.5 mm

132 Quartz Arenite with crossed polars. Locality: New Red Sandstone, Permo-Trias, England.

133

1 mm

133 Quartz arenite with crossed polars. Locality: Cribbath, Wales.

Feldspars in sedimentary rocks

134 and **135** show a coarse-grained sandstone made largely of feldspar and quartz. These minerals are not easily distinguished in the view taken with plane-polarized light (**134**) because they have similar relief. With crossed polars (**135**) three different feldspars can be seen. The large grain near the centre of the picture shows a perthitic intergrowth of two feldspars (see p. 50). To the right and slightly below this is a grain showing the cross-hatched or tartan twinning of microcline, and the smaller grain to the right and below the microcline shows the albite twinning of plagioclase feldspar. On the whole, alkali feldspars are more common in sedimentary rocks than plagioclase because they are more resistant to chemical weathering.

134 Sandstone in plane-polarized light.

135 Sandstone in crossed polars.

Sub-arkose

136 and **137** show a slightly porous sedimentary rock in which the pores have been impregnated with a blue dye-stained resin. Most of the grains are quartz, clear in plane-polarized light (**136**), and shades of grey with crossed polars (**137**). Many show uneven interference colours especially at the bottom of the figure. This indicates that the grains would go into extinction over a range of several degrees rather than all at once. This phenomenon, known as *undulose extinction*, is found in quartz from many igneous and metamorphic sources. The sediment also contains significant feldspar. These grains are most easily recognizable in plane-polarized light (**136**) where they are cloudy and show signs of alteration. Partial solution of the feldspar is well shown by the porous areas filled with blue dye. A little clay showing bright interference colours is also present. A quartz-rich sediment, but with more than 5% feldspar classifies as a sub-arkose.

136 Sub-arkose in plane-polarized light. Locality: Millstone Grit, Upper Carboniferous, Northern England.

137 Sub-arkose with crossed polars. Locality: Millstone Grit, Upper Carboniferous, Northern England.

Arkose

Arkoses are sandstones in which more than 25% of the grains are feldspar. The sediment illustrated in **138** and **139** is very feldspar-rich, the feldspars being clearly distinguished from quartz in the plane-polarized light view (**138**) by their cloudy, brownish appearance as a result of alteration during chemical weathering. Quartz with its greater stability is clear and unaltered. Traces of twinning in the feldspar can be seen with crossed polars (e.g. bottom left **139**). The matrix contains opaque minerals, probably iron oxides.

138

138 Arkose in plane-polarized light. Locality: Torridonian, Precambrian, Scotland.

139

139 Arkose with crossed polars. Locality: Torridonian, Precambrian, Scotland.

Sub-litharenite

Litharenites are sandstones containing recognizable rock fragments in addition to individual mineral grains. **140** shows a sandstone in which quartz is the dominant grain type, showing clear in the photograph, but which contains a substantial number of rock fragments. Many are of fine-grained sedimentary or metasedimentary rocks and are dark grey or brownish in colour. Rock fragments in this sediment make up less than 25% of the grains and it would thus be classified as a sub-litharenite.

140 Sub-litharenite in plane-polarized light. Locality: Coal Measures, Upper Carboniferous, Northern England.

Litharenite

141 shows a medium-grained sandstone with a mixture of angular clear and cloudy grains. In the crossed polars view of the rock (**141**) the clear grains are revealed to be quartz and the cloudy grains to be fine-grained probably metamorphic rock fragments. The grey-green areas seen in **142** are the clay mineral chlorite (see p. 42). Since rock fragments are the dominant grain type, this rock classifies as a litharenite.

141

141 Litharenite in plane-polarized light.

142

142 Litharenite in crossed polars.

Wacke/Greywacke

143 is a good example of a greywacke, a sandstone with abundant fine-grained matrix between the grains (brownish-grey in illustration). It is poorly-sorted and most of the grains are quartz. There are, however, some rock fragments (e.g. rounded grain, upper left) which makes this a lithic greywacke according to the classification on page 111.

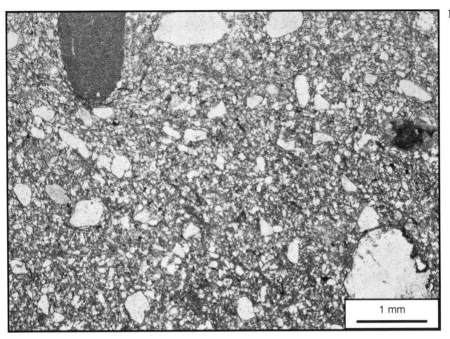

143   Greywacke in plane-polarized light. Locality: Lower Palaeozoic, West Wales.

Micaceous sandstone

144 and **145** show a sandstone with substantial platy muscovite mica, showing bright second-order interference colours in the view taken with crossed polars (**145**). Other grains in the rock are quartz and small fine-grained rock fragments, such that the rock would be classified as a sub-litharenite. However, this takes no account of the mica which would be very evident in hand specimen and lead to the rock being called a micaceous sandstone.

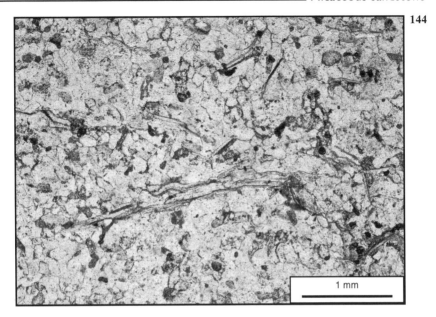

144 Micaceous sandstone in plane-polarized light. Locality: Tilestones, Silurian, Dyfed, Wales.

145 Micaceous sandstone with crossed polars. Locality: Tilestones, Silurian, Dyfed, Wales.

Calcareous sandstone

Illustrated in **146** and **147** is a sandstone which would be classified as a quartz arenite since nearly all the grains are quartz. However, the rock contains substantial calcite both in the form of shell fragments and as a cement between the quartz grains. The calcite can be distinguished by its higher relief – it appears grey in plane-polarized light (**146**) and by its high-order interference colours – mostly pale pinks and greens, seen with crossed polars (**147**). This sedimentary rock may best be described as a calcareous sandstone.

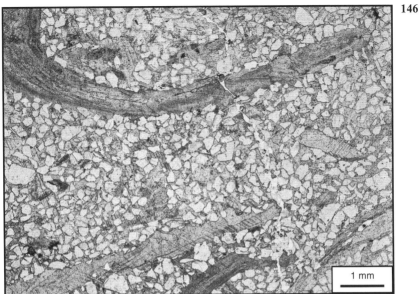

146   Calcareous sandstone in plane-polarized light. Locality: Middle Jurassic, Isle of Skye, Scotland.

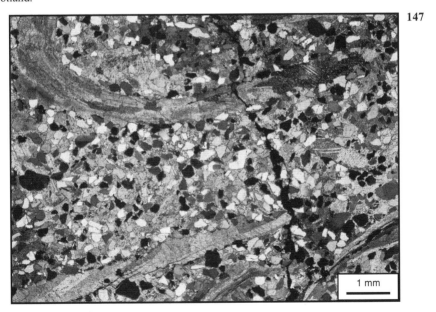

147   Calcareous sandstone with crossed polars. Locality: Middle Jurassic, Isle of Skye, Scotland.

Glauconitic sandstone

Glauconite is a potassium iron alumino-silicate which forms in shallow marine environments and is widespread in sandstones and limestones. **148** and **149** show a glauconitic sandstone cemented by calcite. The glauconite occurs as rounded aggregates of very small crystals with a characteristic green colour. This colour masks the interference colours. Quartz is rounded, clear in plane-polarized light (**148**) and showing shades of grey with crossed polars (**149**). The calcite cement with its high relief shows high-order interference colours.

148 Glauconitic sandstone in plane-polarized light. Locality: Lower Cretaceous, Southern England.

149 Glauconitic sandstone with crossed polars. Locality: Lower Cretaceous, Southern England.

Mudstone

Finer-grained rocks form a large part of the sedimentary record but because of their grain size they can be quite difficult to study with an optical microscope. **151** and **152** show a medium to coarse siltstone with the largest grains about 40 microns in diameter (note scale). These larger grains are quartz and the typical first order grey interference colours (can be rather yellow due to section being slightly thick) can be seen in the picture taken with polars crossed (**152**). Muscovite mica is also evident as clear flakes in the view with plane polarized light (**151**) and showing bright second order interference colours with crossed polars (**152**).

150 shows a higher magnification image of a mudstone showing quartz grains, aligned muscovite flakes, very fine grained material which are likely to be clay minerals and an opaque mineral which is pyrite.

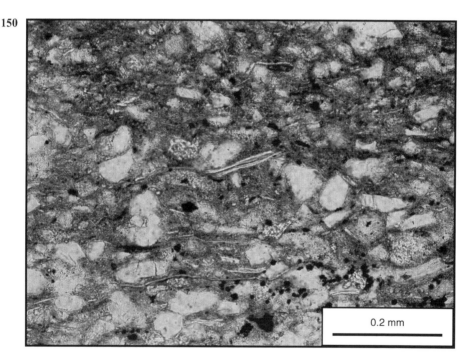

150   Mudstone in plane-polarized light. Locality: Whitby, UK.

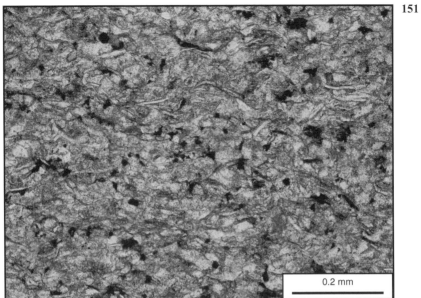

151   Mudstone in plane-polarized light.

152   Mudstone in crossed polars.

Ooid grainstone

Ooids often exhibit a radial structure as well as concentric laminations. A good example is shown in **153**. Nuclei of the ooids are micritic grains and many individual small rounded micritic grains (peloids) are also present, particularly at the top of the figure. Micrite is so fine-grained that individual crystals cannot be distinguished and it appears very dark brownish-grey in thin section. An unusual feature of this sediment is the presence of broken ooids (e.g. left hand side of figure). The rock is grain-supported and is cemented by sparite, seen as clear anhedral crystals filling the spaces between grains.

Ooid packstone

Ooids often show much less well-preserved internal structures than those in **153.** An example of an oolitic limestone in which the ooids are micritic but with faint concentric laminations can be seen in **154.** Ooid nuclei are not easy to make out. Some look like small fragments of curved shell or other bioclasts, but the ooid near the right hand edge about half way down has a quartz grain nucleus with faint concentric laminations. The rock is grain-supported and mostly cemented by sparite. However, carbonate mud matrix is present in small amounts, seen especially in the right hand part of the figure. This makes the limestone a packstone rather than a grainstone according to the Dunham classification (see page 115).

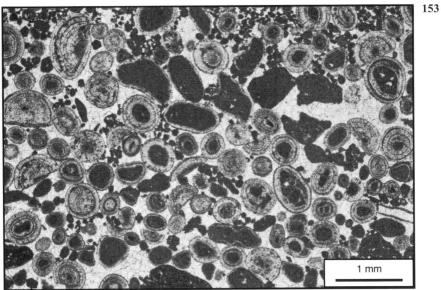

153 Ooid grainstone in plane-polarized light. Locality: Upper Jurassic, Provence, France.

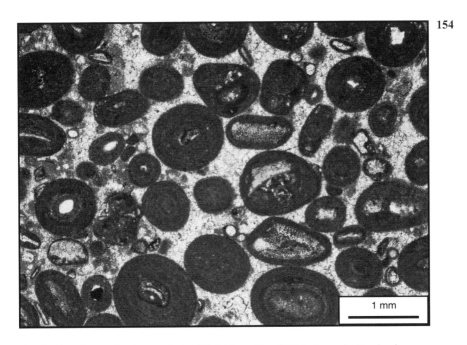

154 Ooid packstone in plane-polarized light. Locality: Middle Jurassic, England.

Bioclast packstone

The sedimentary rock in **155** is a bioclastic limestone with a variety of shell fragments of different sizes. Two principal types of fragment are present; those with a regular, layered structure such as the two large fragments in the lower part of the photograph are shells which have been preserved with their original calcite mineralogy. The smaller fragments such as those in the upper right hand part of the figure are pieces of originally aragonite shell, in which the metastable aragonite has been replaced by calcite sparite. The limestone is grain-supported and contains sparite cement, but carbonate mud matrix is also present and it would thus classify as a packstone.

Bioclast wackestone

The rock in **156** is a matrix-supported limestone. Although there are abundant thin shells and occasional speckled pieces of echinoderm, for example near the top, just left of centre, carbonate mud is the dominant component and the rock thus classifies as a wackestone.

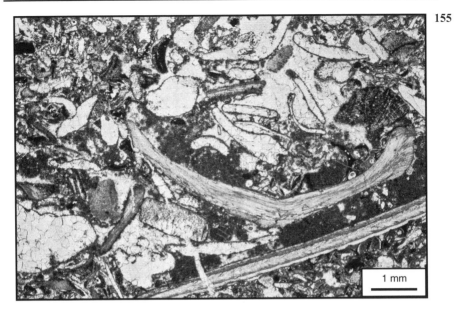

155

155 Bioclast packstone in plane-polarized light. Locality: Middle Jurassic, England.

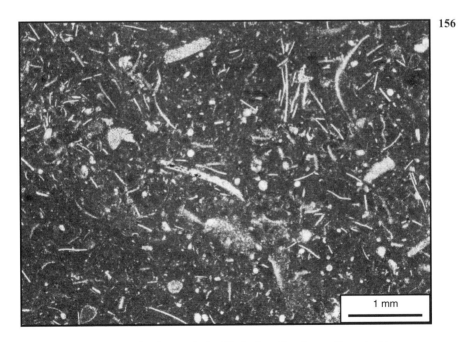

156

156 Bioclast wackestone in plane-polarized light. Locality: Lower Jurassic, Greece.

Intraclast grainstone

Many limestones contain mixtures of different grain types. In **157** probable ooids can be seen (lower left), although internal concentric laminations is are not clearly visible in the photograph, and bioclasts are present (e.g. shell fragments on the right hand side). However, the dominant grains in the field of view are the intraclasts – composite grains in which individual components were deposited, cemented together and then reworked. Intraclasts in this rock contain abundant angular quartz grains, the low relief of the quartz contrasting with the high relief of the sparite cement.

Peloid grainstone

In **158** many of the grains are made of structureless micrite and are thus peloids. A few grains such as the shell fragment in the lower right of the figure are bioclasts with micrite coatings. These are formed by *micritization*—a process involving boring of the shell by microbes and infilling of the borings with micrite. The micrite-filled borings can just be made out at the margin of the shell fragment. To the right of this fragment in the bottom right hand corner of the photograph is a section through a small chambered gastropod shell. Because the original shell wall was made of metastable aragonite, this has since dissolved and the space has later been filled with sparite cement.

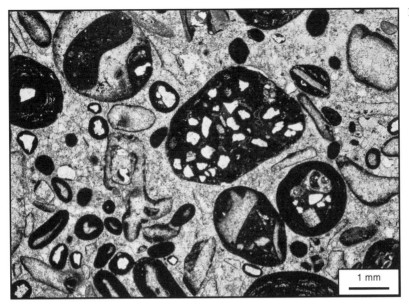

157 Intraclast grainstone in plane-polarized light. Locality: Jurassic, Greece.

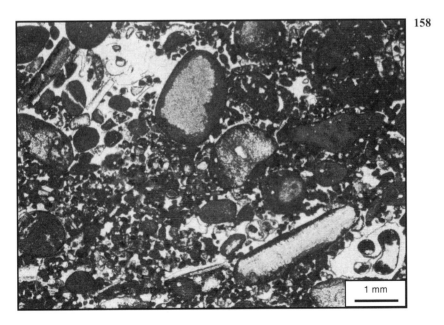

158 Peloid grainstone in plane-polarized light. Locality: Upper Jurassic, Morocco.

Carbonate mudstone

The rock in **159** is almost entirely made of carbonate mud with a few indistinct bioclasts visible. Texturally it is thus a mudstone, using the Dunham classification, since it contains less than 10% grains, but is best called a carbonate mudstone or lime mudstone to distinguish it from a terrigenous clastic mudstone.

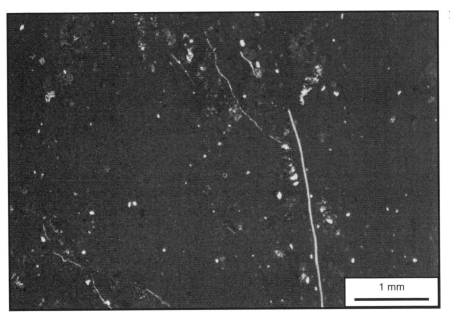

1 mm

159 Carbonate mudstone in plane-polarized light. Locality: Upper Jurassic, Morocco.

Dolomite

Dolomite, $CaMg(CO_3)_2$, is present in many sedimentary carbonate rocks, usually replacing pre-existing calcium carbonate. Optically, dolomite is similar to calcite, but in sediments, dolomite usually occurs as rhombohedra with distinctive rhombic (parallelogram-shaped) cross-sections. Dolomitic carbonate sediments rocks can be classified according to their dolomite content:

0–10% dolomite: limestone
10–50% dolomite: dolomitic limestone
50–90% dolomite: calcitic dolomite
90–100% dolomite: dolomite

The term 'dolomite' is thus used for both the mineral and a rock made up largely of that mineral. This can lead to confusion and the term *dolostone* following the style of the Dunham limestone classification, is sometimes used for the rock.

160 shows a dolomitic limestone. The original sediment is a peloidal packstone, but parts of the rock, especially the matrix have been replaced by dolomite, here showing its characteristic euhedral rhombic shape.

It is not always easy to distinguish dolomite from calcite under the microscope. When individual dolomite crystals growing in a sediment meet one another, continued growth as rhombohedra is not possible and the euhedral shape is lost. A simple chemical technique is often used to help distinguish calcite from dolomite. A thin section is immersed in a solution of a stain called *Alizarin Red S* in weak hydrochloric acid. Calcite reacts with the acid and a reddish-coloured precipitate forms. Dolomite does not react so readily with weak acid and remains unchanged.

161 shows a limestone section which has been treated with Alizarin Red S. It is a dolomitic limestone, calcite showing shades of pink, red and brown. In this case the shell fragments are pink and the micritic intraclasts dark red-brown. The dolomite is colourless to grey and appears to be preferentially replacing matrix and/or cement.

162 shows a calcitic dolomite. Euhedral rhombic crystals of dolomite make up most of the rock, but part of the original limestone remains. The limestone is comprised of dark-looking fine-grained carbonate mud between the dolomite crystals. Note that, although many rhombic crystals can be clearly seen, where the crystals have grown together, such as in the left hand side of the field of view, the rhombic shape is lost.

A dolomite rock (dolostone) can be seen in **177**.

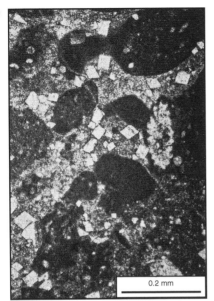

160

161

160 Dolomitic limestone in plane-polarized light. Locality: Jurassic, Greece.

161 Dolomitic limestone treated with Alizarin Red S in plane-polarized light. Locality: Carboniferous Limestone, South Wales.

162 Porous calcitic dolomite in plane-polarized light

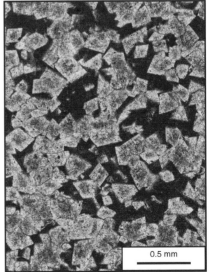

162

Radiolarian chert

Cherts are rocks composed of *authigenic* silica – that is silica formed either by precipitation from water or as a secondary mineral within the sediment. Silica is usually in the form of fine-grained quartz. Primary cherts comprise mostly the remains of organisms which secrete siliceous hard parts such as some sponges and the microfossils radiolaria and diatoms. **163** and **164** show a radiolarian chert. The sample shows the spherical radiolarian tests and a few thin spines set in a matrix masked by red-brown iron oxide. The fine-grained nature of the quartz making up the radiolaria is evident in the crossed polars view (**164**).

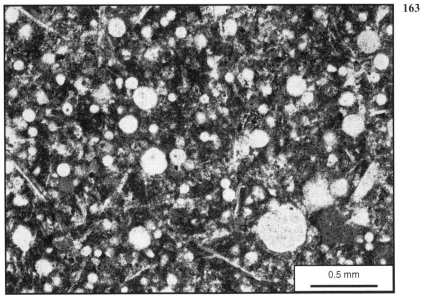

163 Radiolarian Chert in plane-polarized light. Locality: Lower Cretaceous, Greece.

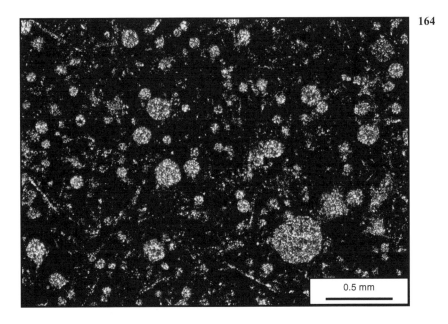

164 Radiolarian Chert with crossed polars. Locality: Lower Cretaceous, Greece.

Replacement chert

Many cherts are secondary, with silica usually replacing limestone. Replacement is often partial, with silica preferentially picking out certain shell fragments, but may develop into nodules or layers. **165** and **166** show a chert from a layer within a carbonate rock succession. The nature of the limestone before replacement is suggested by the round to elliptical grain sections (peloids or ooids?) and the long narrow grains (shell fragments?). Most of the original grains have been replaced by fine-grained quartz. The brownish areas in the centre of the figure are made of quartz showing a fibrous structure (seen with crossed polars in **166**). This is a variety of silica known as *chalcedony* and is probably a pore-fill rather than a replacement. The large grain of quartz at the left-hand edge of the figure is a detrital grain.

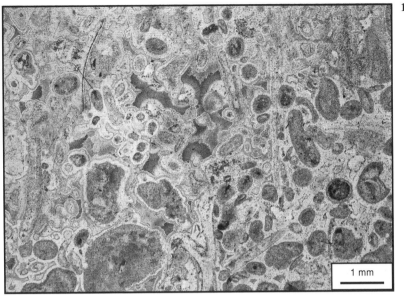

165 Replacement Chert in plane-polarized light. Locality: Upper Jurassic, southern England.

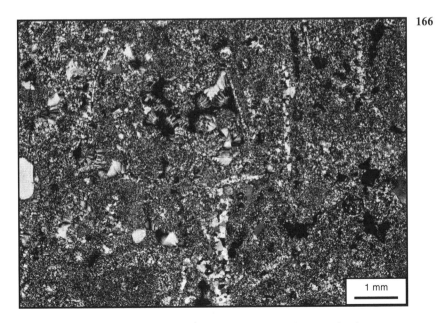

166 Chert with crossed polars. Locality: Upper Jurassic, southern England.

Evaporite

Evaporite minerals are those which precipitate from natural waters concentrated by evaporation. Only a few minerals are common in evaporite deposits, but because of their solubility the minerals are particularly susceptible to changes during diagenesis and complex textures may result. **167** and **168** illustrate a marine evaporite comprising two minerals, *halite* and *anhydrite*. Halite is rock salt, NaCl, and is isotropic. It forms the low relief layers in plane-polarized light (**167**) which are black with crossed polars (**168**). Anhydrite, $CaSO_4$, shows moderate relief and bright mostly second order interference colours. In this sample most of the anhydrite is fine-grained but some characteristic rectangular crystals are also well seen.

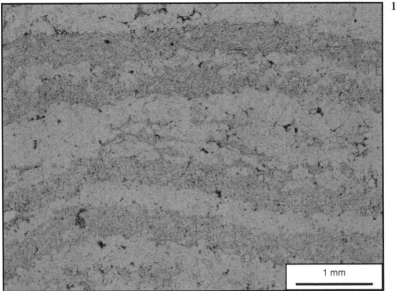

167 Halite and anhydrite in plane-polarized light. Locality: Permian, northeast England.

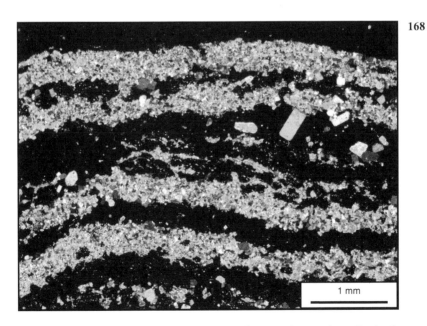

168 Halite and anhydrite with crossed polars. Locality: Permian, northeast England.

Ooidal ironstone

Iron minerals are present in small amounts in many sedimentary rocks. Occasionally, there is sufficient iron for the rock to be of economic value as iron ore. Such rocks are called sedimentary ironstones. One type of ironstone well-known, for example, in the Jurassic rocks of Europe has abundant carbonate and textures similar to limestones with ooids and shell fragments present. Ooids may be made of iron oxides or silicates. In **169** and **170** ooids consist of *berthierine*, an iron alumino-silicate. In this example berthierine is pale brown although commonly it is green. It is recognizable by its very low birefringence, seen in the crossed polars view (**170**) as almost black. The brownish, high relief crystals with high birefringence replacing the margins of some of the grains are *siderite*, $FeCO_3$, and a clear calcite cement is present.

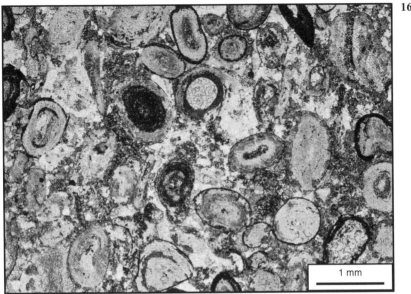

169 Ooidal ironstone in plane-polarized light. Locality: Lower Jurassic, Britain.

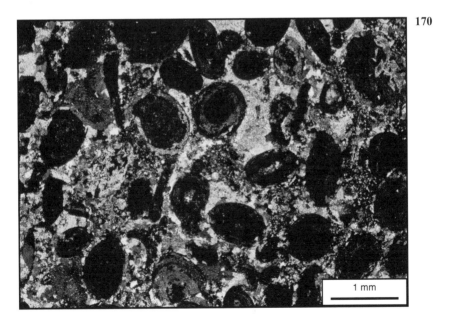

170 Ooidal ironstone with crossed polars. Locality: Lower Jurassic, Britain.

Banded ironstone

Precambrian iron formations are distinctly layered and are often known as Banded Ironstones. **171** and **172** show iron-oxide rich layers (opaque) and fine quartz showing its characteristic first-order interference colours with crossed polars (**172**).

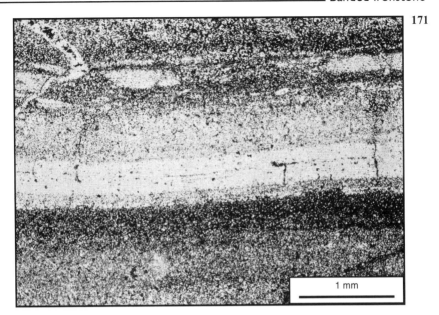

171

171 Banded ironstone in plane-polarized light. Locality: Precambrian, Transvaal.

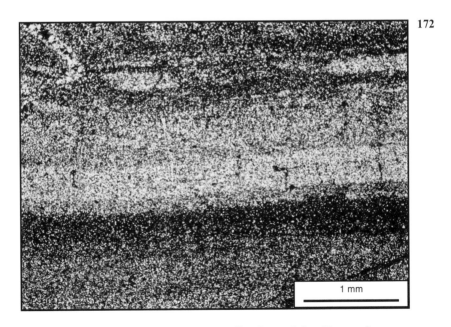

172

172 Banded ironstone with crossed polars. Locality: Precambrian, Transvaal.

Volcaniclastic rocks

Pyroclastic rocks are clastic rocks composed primarily of volcanic material. Where this volcanic material has been transported and reworked through mechanical action, and lithified, the rocks are called volcaniclastic.

They are a diverse and difficult group of rocks sometimes studied with sediments as many have been reworked by sedimentary processes, but more often treated with igneous rocks.

173 shows a volcanic conglomerate in which the fragments are of basaltic composition. Such a rock with recognizable rock fragments could be classified according to the normal sandstone classification (**128**) as a litharenite, although in this field of view fragments greater than 2mm in diameter are dominant and hence the rock is better called a conglomerate (or with pyroclastic terminology, an agglomerate).

174 and **175** show a rock composed of large subhedral feldspar crystals and, mostly smaller, curved or elongate fragments of volcanic glass (isotropic and hence black with crossed polars, **175**). A rock composed of airfall material of volcanic origin and where the pyroclast size is less than 2 mm is known as a *tuff*. A **lithic tuff** contains a predominance of rock fragments, a **vitric tuff** contains a predominance of pumice and glass fragments, and a **crystal tuff** contains a predominance of crystal fragments. Hence this example could be called a crystal-rich vitric tuff. Tuffs deposited subaerially and in which the fragments are not fully cooled may be modified to welded tuffs or ignimbrites, figured under igneous rocks (**126, 127**). In **174** and **175** a carbonate matrix showing high-order interference colours is also visible.

173

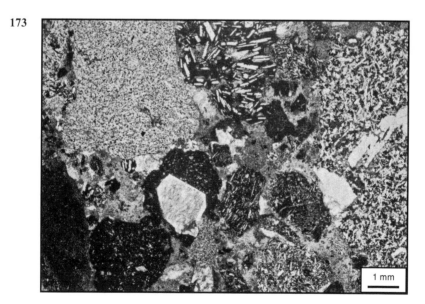

173   Volcanic conglomerate of basaltic fragments in plane-polarized light. Locality: Quaternary, Réunion.

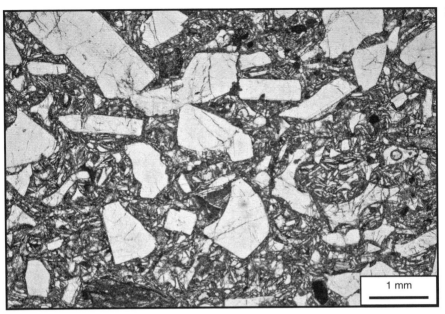

174

174 Crystal tuff in plane-polarized light. Locality: Miocene, Mallorca.

175

175 Crystal tuff with crossed polars. Locality: Miocene, Mallorca.

Porosity in sedimentary rocks

Porosity types in sedimentary rocks have been introduced on page 110. Two examples of porous carbonate rocks are shown here. In both these examples the rock sample was impregnated with a blue-dye stained resin before sectioning so that the porosity is clearly seen in a plane-polarised light view. Both these examples also come from the subsurface and have been collected from borehole cores taken in connection with exploration for hydrocarbons.

176 is a very porous limestone from the Arab D Formation in Saudi Arabia. This unit forms the world's biggest oil reservoir. The limestone is a grainstone with a mixture of bioclasts, intraclasts and peloids. There is a high intergranular porosity between the grains and some intragranular porosity within Foraminifera (chambered microfossils seen, for example, on left hand side of photograph). Some of the pore-spaces between grains look quite large and it is possible that a component of the rock has dissolved out to increase overall porosity.

177 is a dolomite rock, or dolostone (p 146), with rhombic crystals well seen. The different shades of grey shown by the dolomite result from its variable relief depending on orientation (as with calcite, p 58). The porosity type shown here between the dolomite rhombs is called intercrystal porosity.

Porosity can also be seen in **131** and **132** (intergranular porosity in a quartz arenite) and in **136** and **137** (solutional porosity in a sub-arkose).

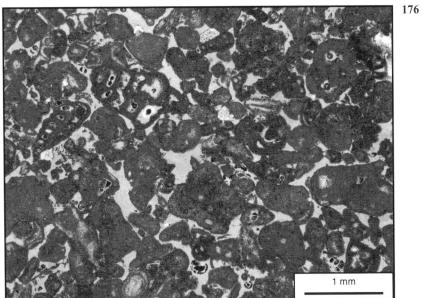

176

176   Limestone in plane-polarized light. Locality: Arab D Formation in Saudi Arabia.

177

177   Dolomite in plane-polarised light.

PART 5

Metamorphic rocks

Metamorphism is the name given to the process or processes by which the mineralogy and/or texture of a pre-existing rock is changed. The rock may have originally been igneous, sedimentary, or even a previously metamorphosed rock, before being subjected to a change in physical conditions such that the mineralogy or texture is altered. It is generally assumed that the bulk chemical composition of the rock is largely unchanged except for the loss, or sometimes gain, of volatile constituents such as water or carbon dioxide.

Where it can be demonstrated that there has been a significant gain or loss of non-volatile constituents on a scale much greater than the size of a thin section or a hand specimen, the term *metasomatism* instead of metamorphism is used. That metasomatism has taken place on a large scale has been established for some rocks but we shall not consider this possibility further here because, as noted in the preface, we are concerned more with the description of rocks than their origins.

The presence in many metamorphic rocks of minerals which only form at high pressures and temperatures indicate that the rocks have, in many cases, been buried to considerable depth in the crust and have subsequently returned to the surface. In addition, structural and textural features of many metamorphic rocks indicate that they have also undergone deformation. The main agents of metamorphism are therefore temperature and overburden (or lithostatic) pressure which control the mineral assemblages produced. *Thermal* or *contact metamorphism* is the term given to the process in which the main agent is increase in temperature caused by the intrusion of an igneous mass. A thermal or contact *aureole* is formed in the rocks surrounding an igneous intrusion, the metamorphic effect decreasing outwards from the igneous mass. The term *hornfels* is commonly used for rocks metamorphosed in this way, particularly those formed at the highest temperature. *Dynamic metamorphism* is caused by movement along a major fault or thrust. The rocks may undergo deformation, reducing the grain size, by either brittle, ductile or plastic processes, to form finer grained zones within intact rocks. These fault zones vary in thickness from mm to km's. The fault rocks formed in this way are known as *cataclasites* or *mylonites*.

Where intrusions occur deeper in the crust where the temperature of the host rocks are already high, contact metamorphism grades into low pressure/high temperature regional metamorphism.

By far the most abundant metamorphic rocks have been formed by *regional metamorphism*. This type of metamorphism is due to large-scale deformation of the crust of the earth at elevated pressure and temperature; the extent to which the rock is changed from its original state is described by the term *grade*. Thus low grade metamorphic rocks show the first signs of change of mineralogy and this begins under conditions where diagenesis of sedimentary rocks finishes – an indefinite boundary. The highest grade of metamorphism passes into the realm of igneous processes – again a very ill-defined boundary – where the formation of a magma or melting of the original rock has occurred on a significant scale.

It was as a result of studies extending from those of large-scale structures down to details of mineral compositions and textures that a picture of the pro-cesses involved in metamorphism was built up. Subsequently, it has been possible to synthesize many of the minerals characteristic of different metamorphic condi-tions and so to assign approximate pressures and temperatures to the conditions of formation of the rocks. Thermodynamic calculations have allowed extrapolation of this data and now mineral paragenesis and mineral compositions can be used to determine detailed pressure and temperature estimates for many metamorphic rocks. Compositional zoning in some minerals and evidence of reaction textures can help constrain pressure-temperature paths which rocks have followed during parts of their history.

The reader is again reminded that although we are here concerned only with the petrography of rocks, it is useful to have a framework by which we can classify them and this necessitates some familiarity with current ideas of the circumstances in which the rocks are formed.

Metamorphic rocks are generally classified according to the *metamorphic facies* to which they belong. This concept was introduced to group together rocks which had been subjected to certain conditions of pressure and temperature irrespective of their bulk chemical composition. The names proposed for each facies were derived from the mineralogy which would be expected from metamorphism of rocks of basaltic composition. At the time the proposal was made, limits of pres-sure and temperature could not be assigned to the different facies and this is still true at the present time, although there is general agreement as to the approximate ranges of temperatures and pressures covered by many of the facies. **178** shows the fields of stability of the metamorphic facies illustrated in this book. Metamorphic facies series are used to describe the variation in pressure relative to temperature observed in different tectonic regimes. High P/T facies series are characteristic of subduction zones (blueschist to eclogite facies), medium P/T facies series of continental collisional zones (greenschist to amphibolite facies) and low P/T facies series of island arc grading into contact metamorphism with can be described as very low P/T facies.

The concept of metamorphic facies, as noted earlier, is based on mineralogy of metamorphosed basic rocks. For metamorphosed mudstones (pelites) a more detailed division based on temperature is provided by the '*Barrovian zones*', orig-inally described by Barrow in 1912 from the Dalradian rocks of NW Scotland, which represents a medium P/T facies series. The incoming of a progressive sequence of *index minerals* from chlorite, biotite, garnet, staurolite, kyanite to sillimanite describe zones of increasing temperature.

The concept of *metamorphic grade* (see earlier) is an alternative general classification to metamorphic facies where rocks are described as low, medium or high grade which are approximately equivalent to greenschist, amphibolite or granulite facies.

Metamorphic rocks can be named according to their texture or to their mineralogy, although is some cases it may be more appropriate to use their *protolith* in the name. Metamorphism affects sedimentary and igneous rocks as well as previously metamorphosed rocks. The effects of metamorphism, which is often accompanied by deformation, may not obliterate previous features of the protolith, especially at low metamorphic grades. Original compositional banding is often preserved even at high metamorphic grades. Hence in some cases it is more appropriate to use the prefix meta- followed by the protolith, e.g., metagabbro, meta-arkose.

Most metamorphic rocks are designated by a term which denotes the texture preceded by the names of one or more of the constituent minerals in order of increasing abundance (e.g. garnet- sillimanite schist). These minerals may be indicators of the grade of metamorphism of the rock. The terms used to designate the texture are schist (182–185), gneiss (186–187) and granofels (180–181) (or hornfels (227–228) if the rock is known to be formed by contact metamorphism). Fine grained schists can be termed slates or phyllites. The term granulite has in the past been used both to denote a texture and also a facies – the highest temperature and pressure facies in regional metamorphism. In its original use to describe a texture it meant that the rock is massive and that the mineral grains show no preferred elongation and are generally of uniform size: the term recommended for a rock with such a texture is *granofels* (page 168). In fact the original granulite is a strongly deformed felsic rock now termed a mylonite! Other definitions require the presence of specific minerals, such as feldspar and quartz. In general, it is best to avoid using granulite as a rock type, and to use granulite facies to refer to rocks metamorphosed at high T and P conditions.

In some cases the mineralogy of the rock allows the use of a specific name. Where greater than 75% of a mineral is present, the suffix –ite can be used. For example, quartzite denotes a rock consisting dominantly of quartz; serpentinite (e.g. 259) a rock where serpentine is the dominant mineral. This does not apply for calcite or dolomite where the term Marble is used.

Eclogite (page 218) is the name given to a rock of basaltic composition but where its mineralogy is very different from basalt in that it consists chiefly of a garnet and a clinopyroxene containing sodium and aluminium in significant amounts. It is known that this mineral assemblage is stable only under high pressure conditions but over a range of temperatures.

Amphibolite (page 210) is a metamorphic rock usually of basaltic composition, consisting dominantly of hornblende and plagioclase. This is distinct from an amphibolite facies rock which may not contain any amphibole.

The texture (or microstructure) forms an important part of the nomenclature and interpretation of metamorphic rocks in thin section so some of the most common textural features are illustrated here, in addition to illustration of rock types. When describing textural features it is important to distinguish observations from interpretation. The interpretation of the relationship between deformation and mineral growth due to metamorphic reactions provides key information on the processes affecting the rock and, while some of the textures produced are relatively simple to interpret, others have generated some controversy, the details of which are beyond the scope of this book.

The classification for metamorphic rocks used here follows the IUGS recommendations published in Fettes & Desmons 2007 (Metamorphic rocks: a classification and glossary of terms).

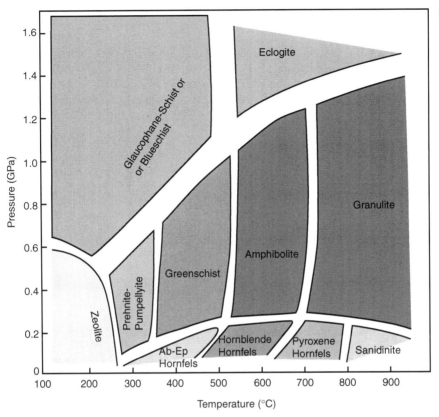

178 Pressure-temperature diagram showing the fields of stability of some metamorphic facies. The diagram is based on Yardley, B.W.D., *An Introduction to Metamorphic Petrology*, Longman, Harlow, (1989).

Textures of metamorphic rocks

Granofels

Granofels is the name given to a massive metamorphic rock which contains no preferred alignment of mineral grains. Compositional banding with layers containing different proportions of minerals can be present (a banded granofels). Fine grained granofels which are found associated with contact metamorphism are commonly called *hornfels*.

The texture of a granofelsic rock is usually either *granoblastic* or *decussate*.

A *granoblastic* texture is a mosaic of equidimensional anhedral grains. Inequant grains, if present, are randomly oriented. The term granoblastic polygonal is used if the grain boundaries are straight. A *decussate* texture is an aggregate of randomly orientated prismatic or elongate crystals, commonly seen, for example, in mica-rich rocks (**179**).

180 & **181** shows a granulite facies pyroxene granofels with a granoblastic texture of intergrown pyroxenes. An opaque mineral and minor colourless, lower relief plagioclase are also present.

179

0.2 mm

179   Mica showing an equilibrium decussate texture in crossed polars.

180

180 Pyroxene granofels with a granoblastic texture in plane-polarized light.

181

181 Pyroxene granofels in crossed polars.

Schistosity and schists

Metamorphism accompanied by deformation produces a schistosity, the oriented or planar structure seen in *schists* due to the parallel alignment of platy or prismatic crystals. The schistosity can be *continuous* (**182 & 183**) where the aligned minerals are evenly distributed through the rock, or *spaced* (**184 & 185**), where the aligned minerals are in regularly spaced zones, separated by layers of more granoblastic minerals (called microlithons). A regular spaced schistosity is often produced as a result of an initial crenulation.

Continuous Schistosity

182 & 183 show a mica-rich schist with a *continuous schistosity* which is defined in plane-polarized light (**182**) by aligned small light green chlorite lath-shaped grains, colourless muscovite and larger dark green biotite porphyroblasts (which contain pleochroic haloes around zircon inclusions). The colourless phase is quartz which has first order grey colours in crossed polars (**181**) where it can be seen that the quartz grains also show some shape elongation along the schistosity. Small higher relief epidote grains, showing brighter colours under cross polars, are also aligned along the schistosity. An opaque phase is also present.

Spaced Schistosity

184 & 185 show a mica schist with a *spaced schistosity* defined by aligned mica, mainly brown biotite, and zones rich in quartz. There are also inclusion-rich albite

182

183

182 & 183 Mica schist in plane-polarized light and crossed polars showing continuous schistosity. Locality: Spean Bridge, N.W. Scotland.

porphyroblasts (poikiloblasts) which are more obvious in crossed polars (**185**), and some small very high relief grains of kyanite. Such a spaced schistosity is normally by formed by recrystallization following crenulation of an earlier schistosity (see page 178).

184　Mica schist in plane-polarized light showing a spaced schistosity.

185　Mica schist in crossed polars showing a spaced schistosity.

Gneissic foliation and gneisses

Under higher temperature conditions of metamorphism and deformation, rocks tend to become coarser grained and develop a gneissic foliation. This is a composite foliation in a medium to coarse-grained metamorphic rock with a compositional banding defined by alternating schistose and granular layers or lenses (found in medium to high grade metamorphic rocks). The rock is called a *gneiss*.

186 & **187** show a quartz-feldspar gneiss derived from deformation and metamorphism of a granitic rock. It contains large alkali feldspars and in hand specimen would be called an *augen gneiss*. Colourless quartz and feldspar make up the rest of the rock with some mica, green biotite and colourless muscovite with high interference colours. Note the slight alteration of the feldspar seen in plane polarized light (**186**) which helps distinguish it from quartz.

In medium to high grade gneiss terrains, heterogeneous areas of rock are commonly present, containing intermixed irregular light patches (leucosomes) which resemble igneous rocks, and dark layers and lenses, observed on a handspecimen or larger scale. Such rocks are called *migmatites* which literally means a 'mixed rock'. Migmatite is a term applied on a hand specimen or larger scale – in thin section these rocks would be termed gneisses.

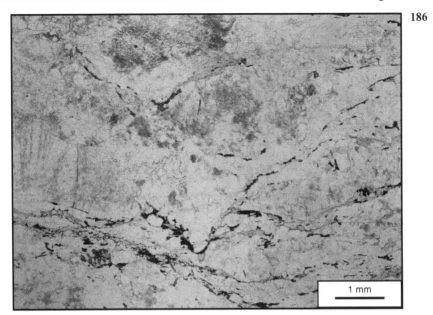

186

186 Granitic gneiss in plane-polarized light. Locality: Austria.

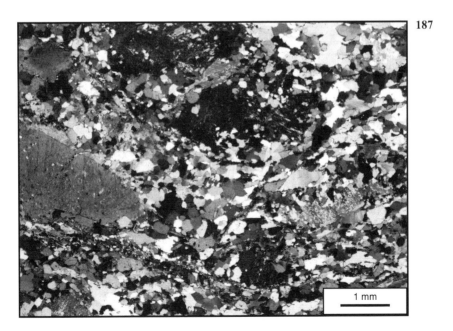

187

187 Granitic gneiss in crossed polars. Locality: Austria.

Mylonite

Mylonites are fault rocks which are cohesive and characterised by a well developed schistosity resulting from intense deformation and tectonic reduction of grain size, and commonly containing rounded to elongated *porphyroclasts* and rock fragments of similar composition to minerals in the matrix. Fine scale layering and an associated mineral or stretching lineation are commonly present. Brittle deformation of some minerals may be present, but deformation is commonly by crystal plasticity. Mylonites may be subdivided according to the relative proportion of finer grained matrix into protomylonite, mesomylonite and ultramylonite.

188 shows a olivine mylonite where the olivine has been deformed and recrystallized to form a fine grained matrix with elongate porphyroclasts. In the centre of the field of view there is also an elongate grain of orthopyroxene showing first order grey interference colours. The black grains are deformed and elongate Cr-spinel.

189 to **190** are from a fault zone cutting a quartz- feldspar rich pegmatite. The quartz has undergone intense plastic deformation and dynamic recrystallization resulting in grain size reduction. Grains in an orientation for easy intracrystalline slip have become elongated, forming a marked mineral stretching lineation seen towards the top of **190** where a single long grain of quartz shows black interference colours. In contrast, the feldspar porphyroclasts (higher relief in ppl and cloudy due to alteration) have behaved in a relatively brittle fashion under the same temperature conditions and grains have fractured and rotated. Single grains of sillimanite and white mica (higher interference colours) have been deformed into mica 'fish', lenticular or lens-shaped grains which are often used to infer the sense of shear in mylonites.

188

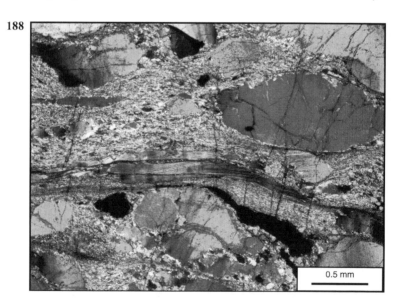

0.5 mm

188   Olivine mylonite in crossed polars. Locality: Val d'Ossola, Ivrea zone, N. Italy.

189

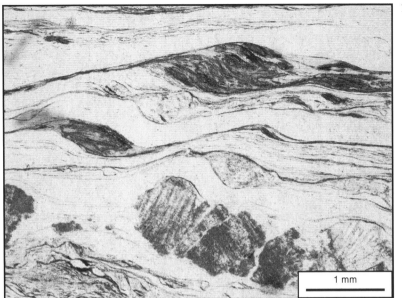

189   Mylonite in plane-polarized light. Locality: Cabo de Creus, SE Spain.

190

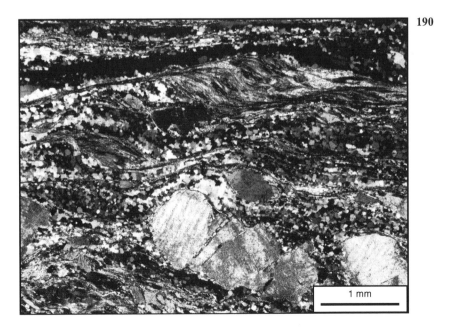

190   Mylonite in crossed polars. Locality: Cabo de Creus, SE Spain.

Cataclasite

A cataclasite is a rock intensely deformed at low temperature by processes involving brittle fracturing and frictional sliding between the fragments. Cataclasis is often localized into fault zones but can be widely distributed. The rock can be cohesive, with a poorly developed or absent schistosity or foliation, or can be incohesive, characterised by generally angular porphyroclasts and rock fragments in a finer-grained matrix of similar composition. Cohesive cataclasites are usually held together either by a cement phase deposited from solution or by clay minerals if present in sufficient proportions. Cataclasites in fault zones are commonly termed fault gouge and are often quite well foliated or banded. Coarse grained cataclasites are often called a breccia.

191 is a fine grained carbonate rock which has undergone brittle deformation and has been re-cemented by coarser grained carbonate infilling the fractures.

192 & 193 show a silicate rock with angular fragments of quartz and feldspar as well as polycrystalline rock fragments in a fine-grained matrix of similar minerals and mica (brighter grains in crossed polarised light).

191

1 mm

191   Carbonate cataclasite in crossed polars. Locality: Monreale, Sicily.

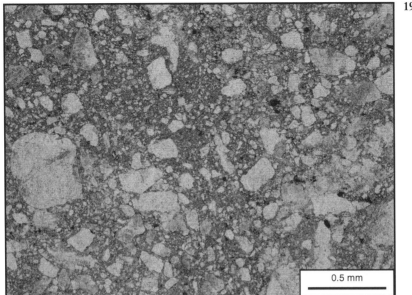

192 Cataclasite in plane-polarized light. Locality: Naxos, Greece.

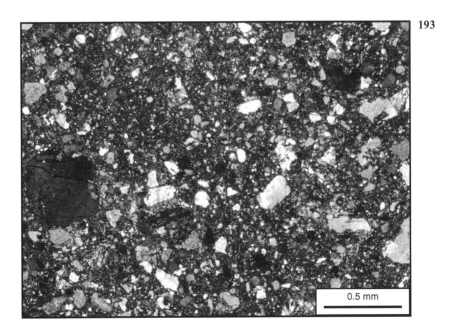

193 Cataclasite in crossed polars. Locality: Naxos, Greece.

Crenulations

A crenulation is formed when a pre-existing foliation in the rock becomes deformed into a series of small folds, eventually forming a new planar feature in the rock parallel to the axial plane of the small folds.

In **194** the foliation defined by aligned mica is crenulated although a new schistosity is not formed. These are chevron folds which occur due to deformation under low temperature (upper crustal) conditions. The minerals are chiefly muscovite, brown biotite and quartz. The clear rounded patches are either garnet (the largest one) or a hole in the slide where the garnet has fallen out in making the section. The difference in the two limbs of the folds is exaggerated by the change in orientation and hence pleochroism of the biotite, one limb having pale yellow biotites and the other showing brown biotites.

In **195** & **196** the crenulations of an earlier foliation (schistosity) are more rounded and deformation is likely to have occurred at slightly higher temperatures, and a new foliation parallel to the fold axial planes is starting to form. The minerals are muscovite and quartz and the black mineral is an opaque mineral. In this case the change of orientation of the muscovite is shown by changes in birefringence colour under crossed polars (**196**). There segregation into mica-rich zones in the limbs and quartz-rich zones in the hinges of the crenulations can also be seen. As metamorphism continues, recrystallization and growth of new minerals along the limbs of the crenulations produces a crenulation cleavage (or schistosity), commonly associated with varying degrees of metamorphic mineral segregation (also known as metamorphic differentiation or transposition of layering). With increasing crystallisation of new minerals, the crenulations can be obliterated and a spaced schistosity produced (see **184** & **185**.)

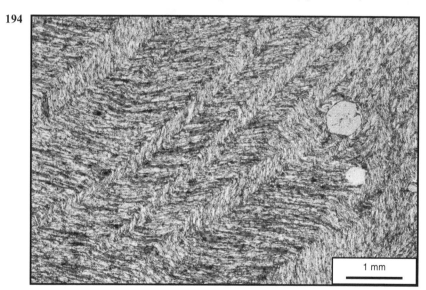

194 Crenulation cleavage in plane-polarized light.

195

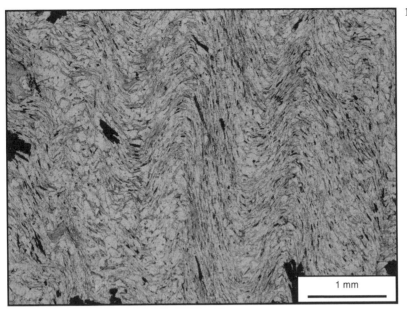

195 Crenulations in a mica schist in plane-polarized light.

196

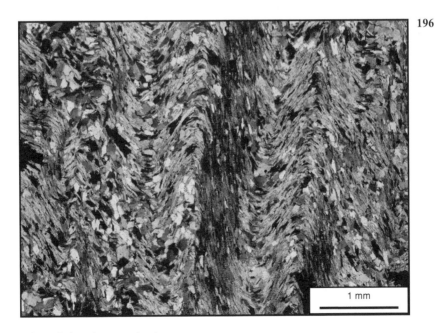

196 Crenulations in crossed polars.

Lineations

As well as planar features, linear structures are common in metamorphic rocks. These can be due to growth of elongate minerals or due to deformation causing elongation of minerals (e.g. quartz in the mylonite **189** & **190**). These structures are most obvious in hand specimen but in thin section they can be seen when minerals which have a shape orientation all display the same crystallographic orientation.

For example, in **197** the majority of the sillimanite grains (see page 196), which make up most of the field of view, display basal sections with a square section, often with cleavage running from corner to corner. They would show low interference colours under crossed-polars. This sample has a strong preferred orientation of needle-shaped sillimanite perpendicular to the section. Brown biotite and an opaque mineral are also present.

Such linear features are common in amphibolites. In **198**, which has been cut perpendicular to the lineation, most amphibole sections show 120° cleavage. These are basal sections show and show browner pleochroism while the elongate grains have one cleavage and are blue/green in colour. The colourless high relief mineral is epidote. In **199** cut parallel to the lineation, all the amphibole grains which are pale green are aligned and elongate with one cleavage. Low relief colourless quartz and plagioclase, colourless high relief phase is epidote, low relief pale green chlorite and an opaque mineral are also present.

197

0.2 mm

197 Lineated gneiss cut perpendicular to lineation showing basal sections of sillimanite, in plane-polarized light.

198

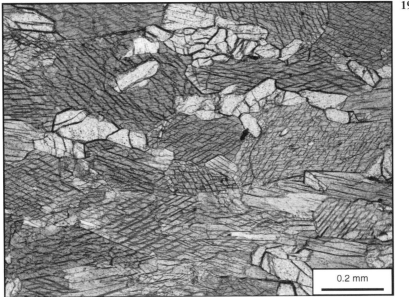

0.2 mm

198   Lineated amphibolite cut perpendicular to the lineation, in plane polarized light.

199

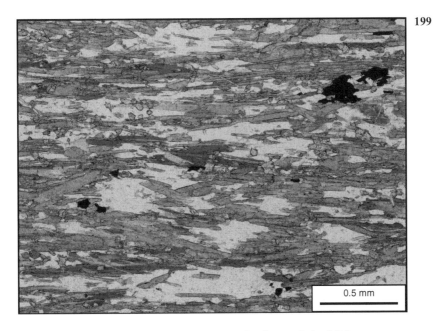

0.5 mm

199   Lineated amphibolite cut parallel to lineation, in plane-polarized light.

Reaction textures

Most metamorphic rocks show mineral parageneses and textures which represent the highest grade of metamorphism to which they have been subjected, but some rocks show evidence of disequilibrium and reactions. The most common form of reaction is alteration of minerals as a result of retrograde reaction under lower metamorphic grade (temperature) conditions, usually involving hydration or carbonation reactions. For example, garnet altering to chlorite or pyroxene to amphibole, feldspar to muscovite (**29**).

Pseudomorphs

Pseudomorphs are volume for volume replacement of one phase of mineral with a distinctive shape by another mineral or minerals, usually as a result of a retrograde reaction. In **200** a euhedral garnet porphyroblast has been partially psudomorphed by fine grained green chlorite.

Corona texture

A corona texture is a zone or zones of contrasting composition formed around a grain as the product of a reaction between the grain and the surrounding mineral(s), forming new minerals.

201 & 202 shows a garnet grain surrounded by a corona of intergrown plagioclase and orthopyroxene (*a symplectite* (see also **203**)) resulting from a decrease in pressure during granulite facies metamorphism. The reaction is: garnet + quartz = orthopyroxene + plagioclase. The orthopyroxene also forms a thin rim corona around the intergrown area. In crossed polars (**202**) it can be seen that the

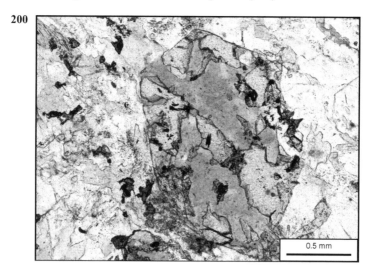

200 Garnet porphyroblast partly pseudomorphed by chlorite in plane-polarized light.

201

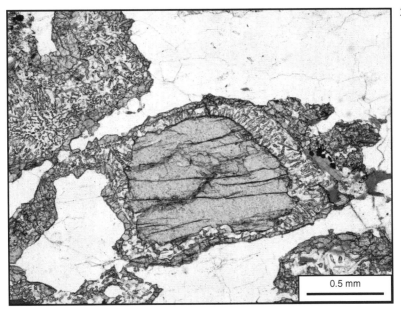

0.5 mm

201 Garnet-pyroxene gneiss in plane-polarized light. Locality: South Harris, Scotland.

202

0.5 mm

202 Garnet-pyroxene gneiss in crossed polars. Locality: South Harris, Scotland.

183

surrounding low relief, colourless phase is quartz. Brown biotite and an opaque mineral are also present. In the top left hand corner, an original garnet has been almost completely replaced by the intergrown reaction products.

Symplectite

This is a fine grained worm-like (vermiform) intergrowth of two or more minerals resulting from a reaction between unstable minerals. When the intergrowth is of quartz in plagioclase the term *myrmekite* is used.

203 is an eclogite where the garnet (high relief, colourless phase in the centre) is no longer stable under lower pressure and temperature conditions and is reacting to produce a fine grained symplectite of green amphibole and colourless plagioclase, and an opaque mineral. Much of the rest of the sample is composed of fine grained symplectites, replacing clinopyroxene and garnet.

This texture indicates that chemical equilibrium in this rock has not been attained and careful study of the minerals should indicate the nature of the reaction. Other textures can be more complex.

Reaction textures

204 is a garnet-kyanite biotite gneiss which also contains corroded grains of staurolite and has fine grained acicular high relief sillimanite growing along grain boundaries. Bioite is the brown platey mineral, the small yellow grains near the left edge are staurolite and the other two high relief phases are garnet (top left) and kyanite which displays a cleavage. The colourless low relief minerals are quartz and feldspar. All these minerals are not stable together but preserve evidence of parts of the high temperature P-T path of this rock. The staurolite is probably relict from an earlier stage in the rocks' P-T path before kyanite- biotite and garnet formed the stable paragenesis. The localised growth of the sillimanite occurred as a result of the partial breakdown of kyanite as the rock has experienced lower pressures during uplift.

0.2 mm

203 Symplectite surrounding a garnet grain in an eclogite in plane-polarized light.

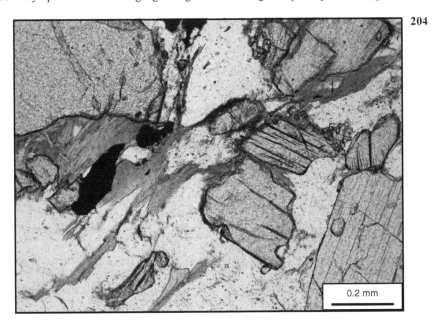

0.2 mm

204 Gneiss showing staurolite and fine acicular sillimanite in plane-polarized light. Locality: Strona-Ceneri zone, N. Italy

Metamorphosed mudstones (pelites)

At low grades of deformation and metamorphism mudstones form into slates and phyllites containing mainly muscovite mica, quartz, chlorite. As temperature (and pressure) increase the rocks become coarser grained and form schists and at higher grades gneisses.

Slates

Under low temperature conditions of metamorphism and deformation, fine grained clay-rich sedimentary and volcanoclastic rocks can form *slates*. A slate has a penetrative cleavage (or schistosity) defined by the preferred orientation of phyllosilicates that are too fine grained to be seen by the unaided eye. In **205 & 206** the bulk of the rock is very fine grained muscovite, quartz and opaque minerals. The cleavage which runs horizontally is defined by aligned micas and needles of opaque mineral, but in crossed polarised light (**206**) it can be seen that many of the mica grains are aligned at a high angle to the cleavage.

Quartz – Muscovite Phyllite

Phyllites (**207** to **209**) are slightly coarser grained than slates so that in hand specimen the muscovite mica produces a shiny schistosity surface. The schistosity is defined by aligned mica, clearer at higher magnification in **209**, as some are very thin, and sometimes the shape elongation of quartz. Detrital quartz grains are often still present (larger grains in **207** to **209**).

Both slates and phyllites can be called fine grained schists. The distinction is clearer in hand specimen rather than in thin section.

205

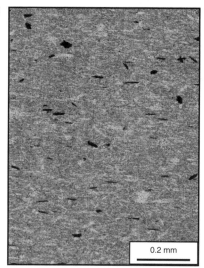

0.2 mm

206

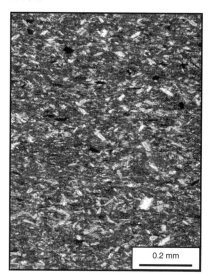

0.2 mm

205 & 206   Slate in plane-polarized light and crossed-polars.

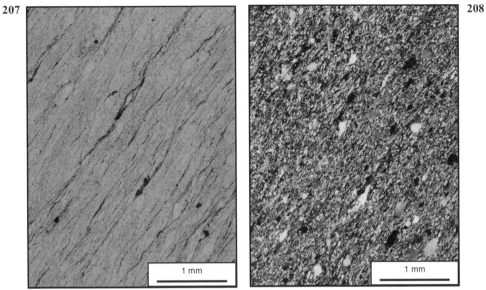

207 & 208 Phyllite in plane-polarized light.

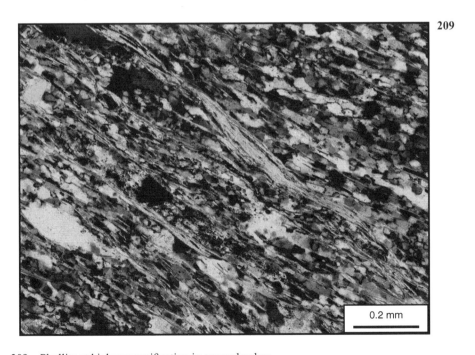

209 Phyllite at higher magnification in crossed polars.

187

Chloritoid schist

The mineral which stand out in relief in the plane-polarized light views of this schist **(210 & 211)** is chloritoid. Chloritoid [$(Fe, Mg)_2 Al_4Si_2O_{10}(OH)_4$] is distinctive by its colour and pleochroism (note some chloritoids have paler colours). In plane polars the chloritoid varies in colour from greenish grey, green to bluish green. With crossed polars **(211 & 213)** the chloritoid is grey to very nearly black.

210

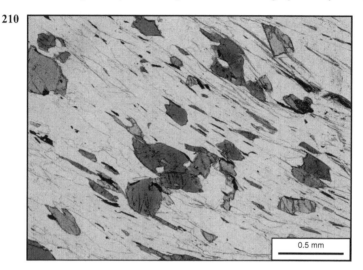

210   Chloritoid mica schist in plane-polarized light.

211

211   Chloritoid mica schist in crossed polars.

There are two reasons for this; the absorption colour is strong and the birefringence in this sample is very low.

The other minerals in these rocks are muscovite, with bright interference colours which is aligned to form the schistosity, and quartz. Chloritoid is found in rocks whose protolith was an Al-rich mudstone.

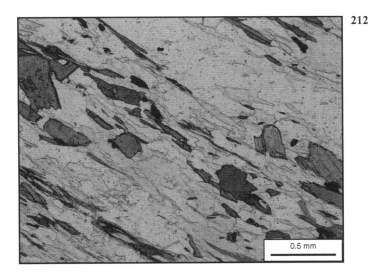

212

212 Chloritoid mica schist in plane-polarized light. Locality: Mallnitz, Austria.

213

213 Chloritoid mica schist in crossed-polars. Locality: Mallnitz, Austria.

Garnet mica schist

This type of rock is formed by medium grade metamorphism of an alumina-rich sediment. The garnets are often seen on the weathered surface of the rock, and in thin section they can be seen without the microscope because of their relief against other minerals.

215 & **216** show euhedral pophyroblasts of high relief garnet in a finer grained matrix of colourless quartz and muscovite, and brown biotite. The rock has a spaced schistosity (page 170) and this can be seen to be a result of a crenulation of an earlier schistosity. The garnets contain inclusion trails which are S-shaped. **214** shows a higher magnification view of these inclusion trails. These are usually interpreted as the garnets having grown at the same time as the deformation which produced the schistosity was occurring (syn-kinematically). Textures such as this can be used to determine the relative timing of deformation and metamorphism, a topic beyond the scope of this book, but covered in many metamorphic petrology books. Plagioclase porphyroblasts, with grey interference colours in crossed polars, can be seen at the top of the field of view in 216.

214

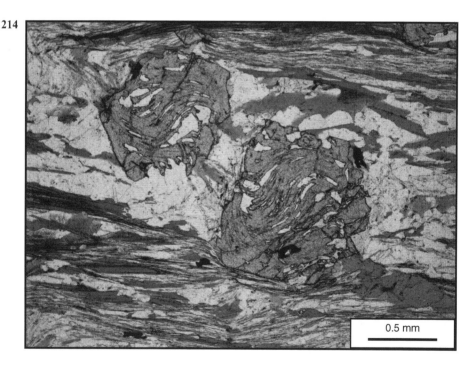

0.5 mm

214 Garnet pophyroblasts in schist in plane-polarized light. Locality: Lukmanier Pass, Switzerland.

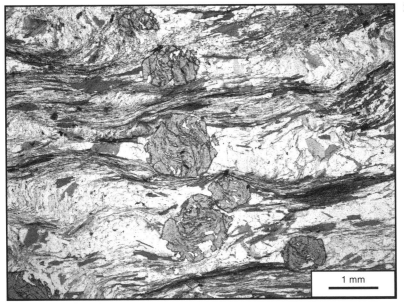

215

215 Garnet mica schist in plane-polarized light. Locality: Lukmanier Pass, Switzerland.

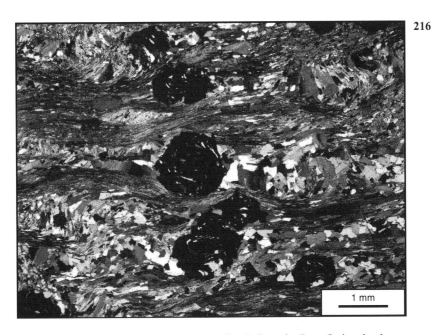

216

216 Garnet mica schist in crossed polars. Locality: Lukmanier Pass, Switzerland.

Staurolite schist

Staurolite ($(Fe,Mg)_2Al_9O_6(SiO_4)_4(O,OH)_2$) is found in mudstones which have been metamorphosed under amphibolite facies (medium grade) conditions. Staurolite shows characteristic yellow pleochroism, and usually occurs as euhedral porphyroblasts. Knee twinning can occur as shown by the large grain in the centre of the micrograph (**218**). In crossed polarised light staurolite shows upper first order to low second order colours (**219**). Muscovite mica, biotite (brown platey minerals), quartz and an opaque mineral are present. The small green mineral is tourmaline, a common accessory mineral in metamorphosed mudstones (pelites).

217 shows euhedral staurolite porphyroblasts in plane polarised light which are full of inclusions (hence a *poikiloblasts*) in a fine grained schist. Irregular brown biotite is also present (partly showing alteration to pale green chlorite) containing dark pleochroic haloes which are characteristic of biotite.

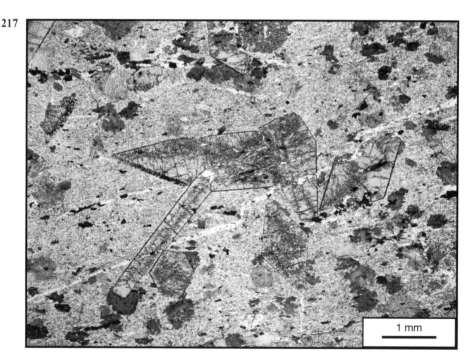

217 Staurolite schist in plane-polarized light.Locality: North Saskatchewan, Canada.

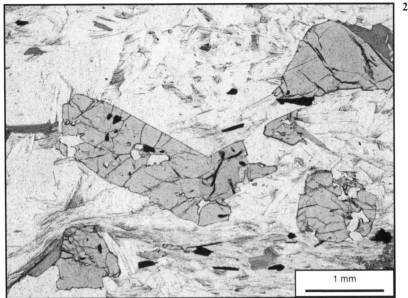

218

218 Staurolite schist in plane-polarized light. Locality: Austrian Alps.

219

219 Staurolite schist in crossed polars. Locality: Austrian Alps.

Kyanite gneiss

These are relatively high grade rock because they contains one of the Al_2SiO_5 polymorphs, kyanite. The high relief colourless mineral is kyanite. It is recognized mainly by its prismatic form, very high relief and most sections show one good cleavage. In **220** where two cleavages are present, they are characteristically at about 80° which helps distinguish it from orthopyroxene where the cleavage would be at 90°. Brown biotite, colourless white mica, quartz and feldspar, and an opaque mineral are also present. Kyanite is also present in **221 & 222** where in crossed polars (**222**) it shows upper first order interference colours (dull yellow to blue). Two large garnets porphyroblasts are also visible in in the top left-hand corner There is also brown biotite and colourless muscovite mica, both showing bright interference colours (**222**) as well as quartz and plagioclase, the latter distinguished by its twinning in crossed polars (**222**).

Kyanite is a very characteristic mineral of metamorphosed alumina-rich sediments and indicates that the highest grade of the amphibolite facies has been attained.

0.2 mm

220 Kyanite showing distinctive cleavage in plane-polarised light. Locality: Glen Urquhart, Scotland.

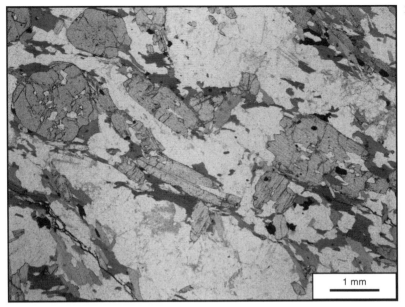

221

1 mm

221 Kyanite gneiss in plane-polarized light. Locality: Glen Urquhart, Scotland.

222

1 mm

222 Kyanite gneiss with crossed polars. Locality: Glen Urquhart, Scotland.

195

Garnet-sillimanite gneiss

Sillimanite is one of the three polymorphs of composition Al_2SiO_5 (the other two are shown in kyanite and andalusite) and is the highest temperature form. It occurs as prismatic crystals, diamond-shaped in cross-section, and also as needle-like acicular crystals. Sillimanite occupies most of the field of view of **223** & **224** where high relief elongate prismatic crystals with no cleavage and middle order orange-blue interference colours are present in the bottom part and square or diamond shaped basal sections, with a characteristic diagonal cleavage (**223**) and low grey interference colours (**224**) are seen in the top part of the image. An opaque mineral and minor feldspar are also seen.

225 & **226** shows a garnet-biotite-sillimanite gneiss where sillimanite is present along with high relief, isotropic garnet, brown biotite (**225**) and colourless lower relief alkali feldspar and plagioclase which show first order grey interference colours.

These rocks has been metamorphosed in the lower pressure part of the upper amphibolite to granulite facies. The presence of alkali feldspar and absence of muscovite indicates it has been metamorphosed at temperatures above the breakdown of muscovite plus quartz.

223

224

223 & 224 Sillimanite gneiss in plane-polarized light and crossed-polars. Locality: South India.

225

225 Garnet-sillimanite gneiss in plane-polarized light. Locality: Val d'Ossola, N. Italy.

226

226 Garnet-sillimanite gneiss in crossed polars. Locality: Val d'Ossola, N. Italy.

Andalusite-cordierite hornfels

It is always wise to look at a thin section of a rock at low magnification first of all and this is particularly true of a rock of this type. With the naked eye it is possible to see that there are white spots in the thin section and before examining the slide under the microscope it is sometimes possible to guess the identity of the minerals making up these spots.

There is in **227** a rectangular region on the left hand side which is clearer than much of the rest of the slide. This area have higher relief than the surrounding minerals and is a euheral andalusite (one of the Al_2SiO_5 polymorphs) porphyroblasts. It also contains a cross- shaped array of minute inclusions and this variety is termed chiastolite. The other oval shaped white patches are *cordierite* [$(Mg,Fe)_2$ $Al_4Si_5O_{18}$]. These cordierite crystals are full of inclusions *(poikiloblasts)*, and some of them show signs of twinning. The grain near the centre of the field of view shows two light grey sectors and two dark grey sectors in the crossed polars view space (**228**). This type of twinning, called *sector twinning*, is characteristic of cordierite in some rocks. The other minerals forming the groundmass of the rock are brown biotite (**227**), muscovite and quartz. Although described as a hornfels, this rock displays a weak foliation (oriented almost diagonally in the micrograph) which has been inherited from the original regionally metamorphosed and deformed slate into which the granite which produced this contact metamorphic rock was intruded.

227

227 Andalusite-cordierite hornfels in plane-polarized light. Locality: Aureole of Skiddaw Granite, England.

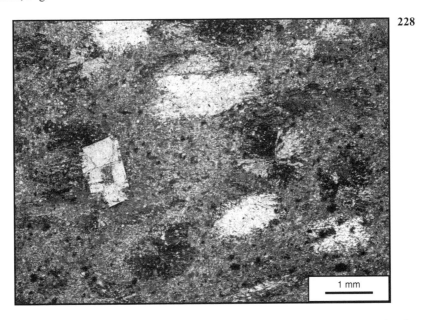

228

228 Andalusite-cordierite hornfels in crossed-polars. Locality: Aureole of Skiddaw Granite, England.

Garnet-cordierite hornfels (Granofels)

Garnet is easily identified in this rock because of its high relief and because it is black in the views with crossed polars (**229** and **231**). Biotite is also fairly readily identified because of its perfect cleavage, brown colour and pleochroism. The rest of the rock is much more difficult. It consists of cordierite along with quartz and feldspar. Cordierite is sometimes mistaken for plagioclase feldspar or quartz because of its low birefringence and low relief. Cordierite is easier to identify when the rock has undergone some late alteration as it becomes pseudomorphed by pinite, a yellowish fine grained intergrowth usually composed of mica and clay minerals. In this rock some cordierite shows lamellar twinning and in this respect also it is very similar to plagioclase feldspar. **229** shows a grain in the centre which shows the sector twinning also seen in **228**. Cordierite is fairly common in alumina-rich rocks which have undergone low pressure or thermal (contact) metamorphism. But note that quarz and plagioclase are much more common minerals!

229

0.2 mm

229 Cordierite in a garnet-cordierite granofels in crossed polars. Locality: Huntly, Scotland.

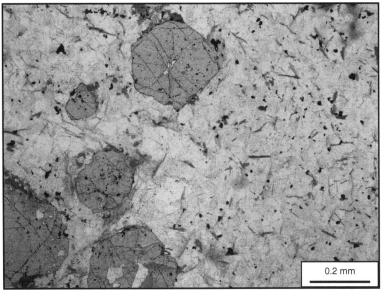

230　Garnet-cordierite granofels in plane-polarized light. Locality: Huntly, Scotland

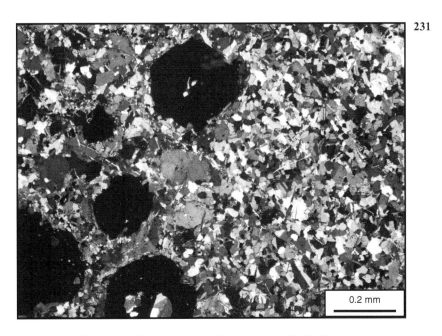

231　Garnet-cordierite granofels in crossed polars. Locality: Huntly, Scotland.

Metamorphosed carbonate rocks

Metamorphosed limestones composed of calcite recrystallize to calcite marbles. If the original limestone contained some quartz, at high temperatures calcite and quartz react to produce wollastonite. If the original carbonate rock also contained dolomite, a range of Ca-Mg silicate minerals are produced, some of which are illustrated here.

Tremolite Marble

233 & 234 shows large colourless porphyroblasts (can be termed poiiloblasts due to abundant small inclusions) of euhedral tremolite (Ca amphibole) in a finer grained matrix of carbonate. Much of this carbonate is calcite but dolomite is also present. These can be difficult to distinguish in metamorphic rocks. The best optical method of distinguishing the two relies on the different orientation of lamellar-twin composition planes. Find a grain showing two sets of sharp twin lamellae. The acute angle between the two sets of lamellae is ca. 45° in calcite and ca. 80° in dolomite.

The matrix of this rock shows an equilibrium granoblastic texture (**232**). The twinning shows up in plane polarised light (**232 & 233**) in carbonate minerals as the relief varies with orientation. Note that all the grains in **232** are carbonates with the relief varying depending on orientation.

232

0.2 mm

232 Marble showing granoblastic texture in plane-polarized light. Locality: Campo Lungo, Switzerland.

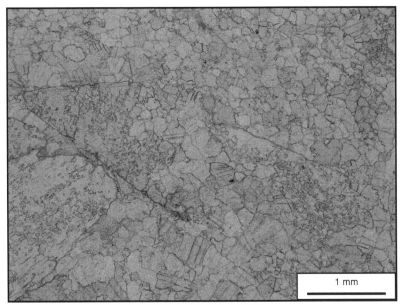

233 Tremolite marble in plane-polarized light. Locality: Campo Lungo, Switzerland.

234 Tremolite marble in crossed polars. Locality: Campo Lungo, Switzerland.

Forsterite and diopside marble

235 & **236** show a marble containing Forsterite (the Mg olivine) as well as calcite. The olivine has high relief and bright interference colours apart from in basal sections which have lower colours. It is characterised by irregular fractures and in this sample there is a small amount of alternation to serpentine (fibrous, low relief, grey interference colours) along grain boundaries and fractures.

237 & **238** show a marble composed mainly of calcite but also contains *diopside* (a clinopyroxene) and brown biotite mica. The diopside crystal are rounded,

235

235   Forsterite marble in plane-polarized light.

236

236   Forsterite marble in crossed polars.

show bright interference colour in the crossed polars view (**238**), and many have a cleavage which distinguishes them from olivine.

The minerals in metamorphosed limestones are often difficult to identify because they are usually colourless, and in the absence of feldspar and quartz it is difficult to judge the correct thickness of the section. The presence of Mg-Si minerals such as forsterite and diopside with the calcite indicates that the original rocks were a siliceous dolomitic limestone.

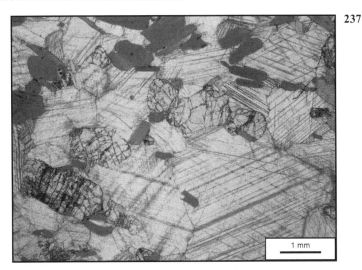

237   Diopside-biotite marble in plane-polarized light.

238   Diopside biotite marble in crossed polars.

Metamorphosed basic rocks

Epidote-actinolite-chlorite Schist (a Greenschist)

240 & **241** show a schist of basic composition (originally a basaltic lava, tuff or dyke) which has been metamorphosed to the greenschist facies. It contains epidote [(Ca$_2$Al$_2$Fe(Si$_2$O$_7$)(SiO$_4$)O (OH)] which is the small, high relief, lozenge shaped grains with a light yellow/brown colour with bright interference colours (**241**). These can be anomalous (do not appear on the colour chart). Two green minerals are present. The larger grains are an amphibole, actinolite, with higher relief which forms porphyroblasts in this rock. It is pleochroic and the colours vary between green and yellow/brown. The interference colours are low second order, cream to pale to mid brown (**240**). Low relief platy pale green chlorite forms much of the matrix and has anomalous grey-white interference colours in crossed polars. The alignment of the chlorite defines the schistosity. The colourless low relief minerals, with low interference colours are quartz and albite which have similar relief and can be difficult to distinguish as the latter often does not show much twinning in greenschist facies rocks. An opaque Fe oxide is present and also an accessory mineral, sphene which has very high relief and extreme pale interference colours but is difficult to distinguish from the epidote at this magnification.

239 shows a similar actinolite-epidote-chlorite schist with some dark green biotite and albite which forms most of the colourless low relief areas. Some of the albite grains contain fine aligned inclusion trails at a high angle to the schistosity which may reflect an original depositional fabric in this rock.

239
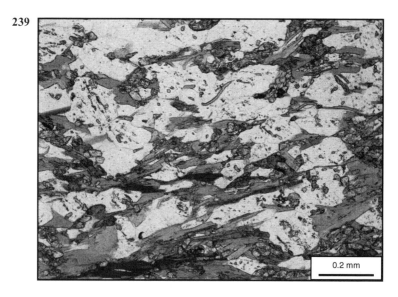

0.2 mm

239 Epidote-actinolite-chlorite schist in plane-polarized light. Location: Islay, N.W. Scotland.

240

240 Actinolite-epidote-chlorite schist in plane-polarized light.

241

241 Actinolite-epidote-chlorite schist in crossed-polars.

Garnet epidote amphibolite

242 can be called an amphibolite as it is composed dominantly of amphibole and plagioclase. It contains pale green- brown pleochroic amphibole (note basal sections with 120° cleavage show pale brown pleochroism while elongate sections with one cleavage are pale green/blue). Eudehral porphrobalst of garnet show high relief and are isotropic under crossed polars. The finer grained colourless, relatively high relief phase with bright interference colours is epidote. Minor low relief pale green chlorite with grey anomalous interference colours (**243**) is present, partly replacing the lower garnet and represents a product of later alteration. This rock is texturally a garnet-epidote-amphibole schist, and metamorphosed in the epidote amphibolite facies. Some of the amphibole grains show blue pleochroism suggesting that this is a relatively high pressure rock which is also consistent with the lack of plagioclase.

242   Garnet epidote amphibolite in plane-polarized light.

243   Garnet epidote amphibolite in crossed-polars.

Amphibolite

Most amphibolites contain amphibole and plagioclase but with increasing meta-morphic grade the amphibole becomes brown in colour and the plagioclase more calcic, and grain size also increases. With increasing grade the amphibole changes from actinolite in the greenschist facies to hornblende in the amphibolite facies which tends to be more equant.

244 is an amphibolite containing green pleochroic amphibole and colourless plagioclase. The colour of the amphibole varies with crystallographic orientation but variation in pleochroic colour within individual grains indicate that there is some chemical zoning present in some of the amphiboles. The opaque grains are rimmed by high relief titanite ($CaTiSiO_5$) (also known as sphene) indicating the opaque minerals must be a Fe-Ti oxide. The alignment of elongate amphibole grains gives this rock a schistosity. This is a typical amphibolite facies rock.

245 & 246 show an amphibolite containing brown pleochroic hornblende and Ca plagiclase which shows distinctive twinning in crossed polars (246). The texture is granoblastic. Pale green higher relief clinopyroxene is also present. This rock could also be called a plagioclase hornblende granofels. This rock has a weak foliation produced by the alignment of the slightly elongate grains so strictly might be termed a gneiss.

244

0.5 mm

244 Amphibolite in plane-polarized light.

245

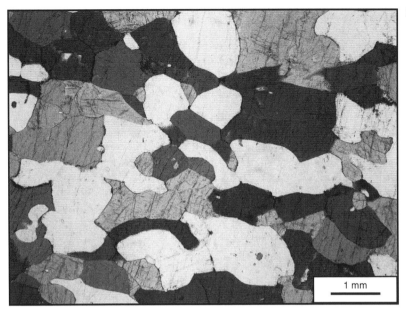

245 Amphibolite in plane polarized light. Locality: Ivrea Zone, Northern Italy.

246

246 Amphibolite in crossed polars. Locality: Ivrea Zone, Northern Italy.

Two pyroxene granofels

247 & 248 shows a granulite facies metabasic rock with an equilibrium texture of intergrown orthopyroxene, clinopyroxene and more strongly coloured green amphibole, along with an opaque mineral. The orthopyroxene shows very pale pink to pale green pleochroism in plane polarised light and has low interference colours while the clinopyroxene is very pale green in plane-polarized light and has intermediate interference colours, blues and reds (**248**).

249 & 250 show another granulite facies rock containing pale pink orthopyroxene, plagioclase and less abundant pale green clinopyroxene which shows higher interference colours. The mineral which shows dark greens and shades of brown (**249**) is an amphibole whose strong colour masks the interference colours. The plagioclase is colourless with low relief and shows distinctive twinning in crossed polars – in this case it is relatively Ca-rich and is of andesine composition.

These rocks show the highest temperature facies in regional metamorphism, and represent granulite facies metabasic rocks.

247

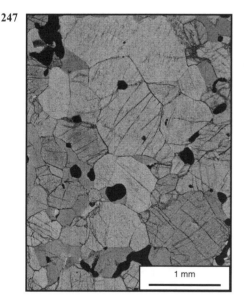

248

247 & 248 Orthopyroxene-clinopyroxene granofels in plane-polarized light and crossed-polars.

249 Pyroxene granofels in plane-polarized light. Locality: North Uist, N.W. Scotland.

250 Pyroxene granofels in crossed polars. Locality: North Uist, N.W. Scotland.

Glaucophane-lawsonite-schist (Blueschist)

The blueschist facies is defined in basic rocks by the coexistence of glaucophane (blue to lilac amphibole) and lawsonite $[CaAl_2(Si_2O_7)(OH)_2.H_2O]$ (stable at high pressure and low temperature) or epidote at higher temperatures. Garnet, quartz, muscovite and jadeitic pyroxene are often present. Schistose rocks with these parageneses, which are blue in hand specimen, are often referred to as blueschists.

251 & **252** contains subhedral lawsonite grains showing distinctive rectangular and diamond shapes, and sometimes showing twinning (grain near centre of **252**), and the blue amphibole, glaucophane which is aligned, defining the schistosity. Most elongate sections show sky blue pleochroism while a few cross section orientations in the top part of the micrograph show 120° cleavage traces and more lavender blue colours. In crossed polars the interference colours are masked to some extent by the colour. The high relief (**251**) accessory mineral is sphene $[CaTiSiO_4(O,OH,F)]$.

251

251   Glaucophane-lawsonite schist in plane-polarized light, Orhaneli, Tavşanlı, NW Turkey.

252

252   Glaucophane-lawsonite schist in crossed polars, Orhaneli, Tavşanlı, NW Turkey.

Glaucophane-epidote (Blue) schist

At higher temperatures, metabasic blueschist facies rocks contain epidote in place of lawsonite. This rock (**253** at lower magnification and **254** & **255**) contains high relief, euhedral pink garnet which is isotropic and hence black in crossed polars, subhedral, high relief epidote (elongate grains, abundant at the top of the field of view, often with darker cores which are likely to be a type of epidote rich in rare earth element, allanite) with bright interference colours, blue pleochroic glaucophane (variable colours due to different orientations), and colourless platey white mica (bright interference colours), and quartz with low relief and low interference colours. The small grains with very high relief are sphene, which has extreme pale interference colours in crossed polars. The schistosity is defined by the alignment of white mica, glaucophane and epidote.

253

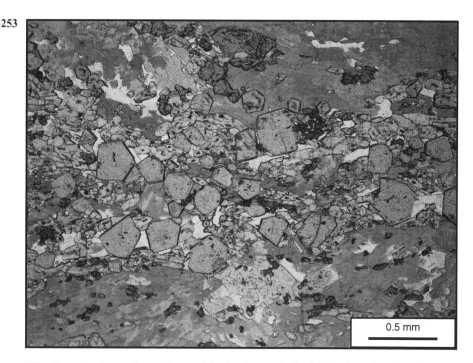

0.5 mm

253 Garnet epidote glaucophane schist in plane-polarized light. Locality: Ward Creek, California, USA.

254　Garnet epidote glaucophane schist in plane-polarized light. Locality: Ward Creek, California, USA.

255　Garnet epidote glaucophane schist in crossed polars. Locality: Ward Creek, California, USA.

217

Eclogite

The two essential minerals in an eclogite facies rock are a magnesium-rich garnet and an omphacitic pyroxene, i.e., a pyroxene which contains sodium and aluminium. Both these minerals require a relatively high pressure for stability and hence the term eclogite is used to name the highest pressure facies.

In **257** & **258** the euhedral garnets are seen most clearly under crossed polars (e.g., top right hand corner) as the other mineral, omphacite, also have relatively high relief. Some white mica (mineral with lower relief and high speckled interference colours adjacent to the garnet) is present along with an opaque mineral.

256 shows a glaucophane- bearing eclogite (blue mineral) along with pale green omphacite and pale pink garnet, altered along fractures to green chlorite, Small elongate epidote grains are present in the matrix and the colourless mineral here is a carbonate mineral showing some twinning. True eclogite facies rocks should not contain plagioclase or biotite.

256 Garnet-glaucophane-omphacite eclogite in plane-polarized light. Locality: Sesia-Lanzo zone, Italy.

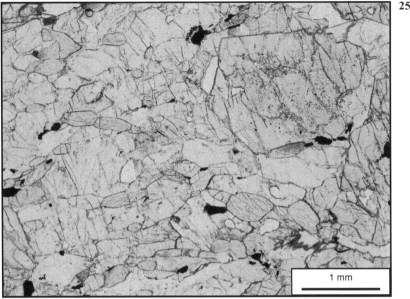

257 Eclogite in plane-polarized light.

258 Eclogite in crossed polars.

Ultramafic rocks (see also peridotites page 72)

Serpentinite

While olivine and pyroxene ultramafic rocks (e.g. 82, 83) are stable under high temperature metamorphic conditions, below ~400°C olivine breaks down to serpentine. **259 & 260** consist almost entirely of the mineral serpentine, hydrated magnesium silicate [$Mg_3(Si_2O_5)(OH)_4$]. The characteristics of serpentine are its low birefringence and the mesh-like texture which can be seen in the view taken with crossed polars (**260**). This form of serpentine shows a pseudomorph texture cut by veins and is formed by hydration and metamorphism of an originally coarse grained olivine-rich peridotite. It is characteristic of the serpentine polymorph *lizardite*.

Serpentinites frequently have relics of olivine and pyroxene crystals from the original igneous rock from which they are formed. In this sample two relics of orthopyroxene are visible, one at the bottom. They are recognized by their high relief and low birefringence compared to the low relief of the serpentine. The section also contains an opaque mineral, probably iron oxide and, on the left hand side dark brown crystals of *spinel* (an oxide of magnesium, iron and chromium) which are isotropic.

259 **260**

1 mm   1 mm

259 & 260 Lizardite Serpentinite in plane-polarized light and crossed-polars. Locality: Lizard Head, Cornwall, England.

Subsequent prograde metamorphism of a serpentinite produces the serpentine polymorph *antigorite,* which characteristically forms a more uniform fine grained decussate texture. Most of **261** & **262** is composed of serpentinite which here forms fine grained laths with low relief and low interference colours, an opaque mineral and a higher relief mineral with higher interference colours (265) which is probably olivine. In **263** the fine grained decussate serpentinite (antigorite) is intergrown with euheral olivine (high interference colours in crossed polars). In this case the olivine is not relict from the original periodotite, but is forming by neomineralization of new olivine grains as a result of reaction during prograde contact metamorphism of the serpentinite.

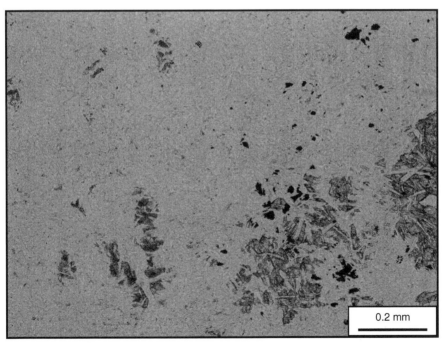

261

0.2 mm

261   Antigorite serpentinite in plane-polarized light. Locality: Val Mallenco, Switzerland.

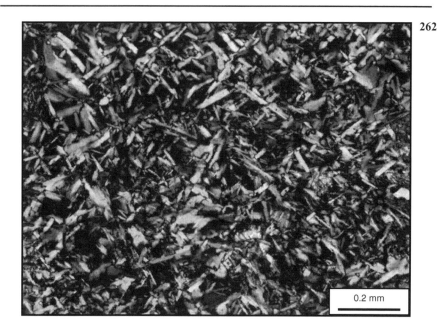

262

262   Antigorite serpentinite in crossed polars. Locality: Val Mallenco, Switzerland.

263

263   Olivine serpentinite in crossed polars. Locality: Val Mallenco, Switzerland.

Appendices

Appendix I

Optical properties of common minerals.

Mineral	Colour	Relief	Cleavage	Shape (ideal)
Quartz	colourless	low	none	
Alkali Feldspar	colourless	low	2 perfect	
Plagioclase	colourless	low	2 perfect	
Olivine	colourless to pale brown	high	1 poor	
Orthopyroxene	grey green to pink, pleochroic	mod. high	2 at 90°	
Clinopyroxene	colourless or pale green or brown	mod. high	2 at 90°	
Amphibole (clino)	variable, colourless to green, blue or brown	mod. high	2 at 60°	
Muscovite (white mica)	colourless	mod. low	perfect basal	
Biotite	green to brown, pleochroic	mod. low	perfect basal	
Chlorite	green to pale straw, pleochroic	low/mod. low	perfect basal	
Calcite	colourless	low to mod high	3 perfect	
Garnet	colourless	high	none	

Max Birefringence	Extinction angle	Twinning	Typical habit/ form	Image no.
first order	0°	none	equant, granoblastic	53-56
first order	0°	simple or tartan	equant, tabular	58-61 186-7
first order	variable	lamellar	equant, tabular	62-4
upper 2nd order	0°	none	equant, rarely prismatic	30-33 188
upper first order	0°	none	prismatic or equant	34-35 249-50
mid 2nd order	~40-45°	simple (rare)	prismatic or equant	37-39, 237-8
mid 2nd order (often masked by colour)	~20°	simple (rare)	prismatic, acicular	40-42 198-9
low 3rd order	0°	v. rare	tabular, platy	46-49 144, 179
upper 3rd order can be masked by colour)	0°	v. rare	tabular, platy	43-45
anomalous (blues, greys, brown)	0°	lamellar (rare)	tabular, platy	50-52 239-241
very high (3rd order)	na	lamellar	equant, granoblastic	69-71 146, 232
Isotropic	na	none	rounded porphyroblasts	72-73 214

Appendix II

Optical properties of additional minerals in metamorphic rocks.

Mineral	Colour	Relief	Cleavage	Shape (ideal)
Sillimanite	colourless	mod. high	1 perfect	
Kyanite	colourless	High	2 at 75°	
Andalusite	colourless or pink	mod. high	2 at 90°	
Staurolite	pleochroic yellow to colourless	high	poor	
Chloritoid	pleochroic: green to slate blue to pale yellow	high	perfect basal	
Cordierite	colourless	low	very poor	
Epidote Group	colourless or weakly pleochroic to pale yellow		1 good	
Serpentine	colourless to pale green	mod. low	perfect basal	
Lawsonite	colourless		2 perfect	
Glaucophane (Amphibole)	Pleochroic blue to greyblue/lilac	mod. high	2 at 60°	

Max Birefringence	Extinction angle	Twinning	Typical habit/form	Image no
low 2nd order	0°	none	prismatic or acicular to fibrous	197, 223-6
upper 1st order	~0.5°	none	prismatic porphyroblasts	220-22
mid 1st order	0°	v. rare	prismatic porphyroblasts	227-8
mid 1st order	0°	knee (rare)	prismatic porphyroblasts	217-19
mid 1st order	0°	simple & lamellar	tabular (often porphyroblasts)	210-213
1st order	na	cyclic & lamellar	granoblastic or rounded porphyroblasts/ poikiloblasts	227-231
1st order (zo, cz) to mid 3rd order (ep). Anomalous	0°	rare, simple in epidote	equant or prismatic	239-241 253-255
mid 1st order	0°	rare	flaky or fibrous	259-263
low 2nd order	0°	simple & lamellar	diamond or rectangular, often porphyroblasts	251-52
mid 2nd order (often masked by colour)	~20°	simple (rare)	prismatic, acicular	253-256

Index

Page numbers are shown in light type, figure numbers in **bold** type.

9781138028067